THE DESIGN OF A PRACTICAL ENTERPRISE
SAFETY MANAGEMENT SYSTEM

The Design of a Practical Enterprise Safety Management System

by

Hossam A. Gabbar

Okayama University,
Okayama, Japan

and

Kazuhiko Suzuki

Okayama University,
Okayama, Japan

KLUWER ACADEMIC PUBLISHERS
DORDRECHT / BOSTON / LONDON

A C.I.P. Catalogue record for this book is available from the Library of Congress.

ISBN 1-4020-2948-9 (HB)
ISBN 1-4020-2949-7 (e-book)

Published by Kluwer Academic Publishers,
P.O. Box 17, 3300 AA Dordrecht, The Netherlands.

Sold and distributed in North, Central and South America
by Kluwer Academic Publishers,
101 Philip Drive, Norwell, MA 02061, U.S.A.

In all other countries, sold and distributed
by Kluwer Academic Publishers,
P.O. Box 322, 3300 AH Dordrecht, The Netherlands.

Printed on acid-free paper

Printed in the Netherlands.

Table of Contents

Table of Figures

List of Tables

Abbreviations

BPCS	Basic Process Control System
CAPE	Computer-aided process engineering
CGU	Control Group Unit
CSC	CAPE-SAFE Controller Common Warehouse Metamodel
CWM	Common Warehouse Metamodel
ETA	Event Tree Analysis
FDS	Fault Detection System
FLD	Fluid
FMEA	Failure Mode and Effect Analysis
FMECA	Failure Mode and Effect Criticality Analysis
FPL	Fault Propagation Layers
FPM	Fault propagation model
FPME	Fault propagation model element
FTA	Fault Tree Analysis
HDE	Hazard Decision Engine
HEF	Hazard Evaluation Follow-up
HEM	Hazard evaluation manager
IPL	Independent Protection Layers
MML	Model Manipulation Language
MMS	Model Management System
MQL	Model Query Language
MUI	Model User Interface
NAS	Normal/Abnormal Situation
OCL	Object Constraint Language
OIS	Operator Interface System
OMG	Object Management Group
OPR	Operation
OSHA	Occupational Safety & Health
P&ID	Piping & Instrumentation Diagram
PEC	Physical Equipment Class

PEEE	Plant Enterprise Engineering Environment
PHA	Preliminary Hazard Analysis
POOM	Plant Object Oriented Model
POS	Position within the process
PV	Process Variable
S/D	Shutdown
S/U	Start-up
SDF	Standard Data Format
SDM	Safety Data Manager
SPS	Safety Procedures Synthesizer
SRM	Safety Regulations Manager
SS	Steady State
STEP	Standard for the Exchange of Product Model Data
UML	Unified Modeling Language
XML	Extensible Markup Language

Glossary

Accident, accident scenario, or accident sequence: An unplanned event or sequence of events that results in undesirable consequences. An incident with specific safety consequences or impacts.

Consequence: The direct, undesirable result of an accident sequence usually involving a fire, explosion, or release of toxic materials. Consequence descriptions may be qualitative or quantitative estimates of the effects of an accident in terms of factors such as health impacts, economic loss, and environment damage.

Event: An occurrence related to equipment performance or human action, or an occurrence external to the system that causes system upset.

Hazard: An inherent physical or chemical characteristic that has the potential for causing harm to people, property, or the environment.

Hazard Evaluation (HE): The analysis of the significance of hazardous situations associated with a process or activity. Uses qualitative techniques to pinpoint weaknesses in he design and operation of facilities that could lead to accidents.

Risk: The combination of the expected frequency (event/year) and consequence (effects/event) of a single accident or a group of accidents.

Risk Assessment: The process by which the results of a risk analysis (i.e. risk estimates) are used to make decisions, either through relative ranking of risk reduction strategies or through comparison with risk targets.

Risk Management: The systematic application of management policies, procedures, and practices to tasks of analyzing, assessing, and controlling risk in order to protect employee, the general public, the environment, and company assets.

Process Operator: A person responsible for monitoring controlling and performing tasks as necessary to accomplish the productive activities of the plant.

Safety System: Equipment and/or procedures designed to limit or terminate an accident sequence, thus mitigating the accident and its consequences.

Simulation: Attempting to predict aspects of the behavior of some system by creating an approximate (mathematical) identifiable model of it. Models might be quantitative, qualitative, or mix.

Preface

This book is a result of research work that has been carried out at the System Analysis Laboratory within the Department of Systems Engineering at Okayama University, Japan. It includes authors' experience in safety management and assessment practices for chemical and production industrial plants. It also includes knowledge and skills gained from collaboration research projects with Loughborough University-UK and VTT-Finland to investigate and design practical enterprise safety management systems as part of computer-aided engineering systems and tools. Safety assessment has been addressed from the process systems engineering viewpoint focusing on health, safety, and environment practices.

Recently, more attention has been made towards health, safety, and environment (HSE) regulations and legislations where the compliance with the national and international safety regulations becomes an essential and deciding factor in the current market competition. Chemical/petrochemical plants are required to apply, comply with, and manage such regulations to keep its leading position in the challenging market. Currently, there is no complete HSE management system that can reliably manage all safety aspects within the enterprise throughout its lifecycle. Such automated system is highly needed to ensure safety regulations and to evaluate the risk associated with the design and operation of any part of the plant process. This comes within the attempts to approach the ideal manufacturing or production plants, and to automate plant enterprise activities while enabling the latest technologies.

This book provides useful materials to enable individual enterprises to develop and implement a complete plant enterprise safety management system, as a part of the HSE management and as integrated with other enterprise management modules in all levels i.e. nano-micromaco. It is not meant to study how to develop information management system, but how to design and implement practical enterprise safety management system as part of the enterprise integrated systems.

The authors will show practical steps to automate and manage all safety aspects within the plant enterprise. The approach used in this study comes in two folds: (1) to develop plant safety model that covers all safety aspects and research the concept of realizing each element; (2) to follow the software engineering lifecycle and to conduct the analysis and user requirements that formulate the contextual and conceptual models to build such integrated safety management system. Drawing the complete picture of the plant enterprise engineering environment helped in understanding the role and links of the proposed solution among other components to automat the different functions within the plant enterprise environment throughout its lifecycle. Object technology is utilized to model and analyze the target system where it is proven to be the most efficient approach to realize robust computer-aided tools.

Acknowledgement

I would like to thank Professor Suzuki for his support and guidance throughout the research progress and development of this book. And I would like to thank Dr. Yukiyasu Shimada for his assistance, and all members of the system analysis laboratory for the interactive discussions. Also, I would like to thank Professor Paul Chung from Loughborough University in UK on his advice and valuable suggestions. And many thanks for participated members from VTT for their valuable comments.

Okayama, Japan.

Hossam A.Gabbar

Part I: INTRODUCTION

1. Overview

1.1. Abstract

Towards the enhancement and improvement of process systems engineering, recent research interests have been focusing on automating different areas within the plant lifecycle. One important area within plant enterprise is enterprise safety. Considerable research efforts are focused on the design and development of practical approaches and successful computer-aided tools and systems that cover different safety functions and activities within plant lifecycle. Although these components could automate and enhance specific safety functions or aspects, however the integration among these components and the complete view is still far beyond expectations. This is simply because most of the developed solutions either focusing on the process safety aspects in the micro level or enterprise risks in the macro level. The integration between these two levels is still unclear and is under investigation. From the system side, most of the proposed automated engineering solutions are more directed towards the management of safety activities within the different functional areas such as design, operation, and maintenance. It is essential to establish integration between the automated engineering modules from the design stage, operational systems, as well as other components within plant enterprise such as human resource, financial and services.

There are different challenges and obstacles to achieve robust enterprise safety management as integrated with enterprise management solutions. The cornerstone of such integrated view is the development of standard and systematic model formalization method that can facilitate the construction of the underlying process design model and build around it other model elements that represent lifecycle activities. Also the model representation mechanism is an essential factor to facilitate the understanding and manipulation (and management) of different model elements throughout plant enterprise lifecycle. This includes operational and management models, which are constructed in different hierarchical levels. In this book, modeling formalization methodology is presented and computer-aided modeling environment is illustrated that enables the accumulation of lifecycle model elements i.e. knowledge, and define and associate plant safety model elements. The proposed modeling methodology will facilitate the analysis stage of the target plant enterprise safety management system where lifecycle model elements will be accumulated in model repository with model query and management functions. UML-based models are developed as part of the requirements analysis for the target enterprise safety management system.

Safety aspects are modeled and represented as associated with plant design model elements. This requires providing robust modeling environment to define and manipulate process design as well as the associated safety model elements. A conceptual architecture of plant enterprise modeling environment (CAPE-ModE) is proposed as a part of plant enterprise engineering environment, called PEEE. This modeling environment will manipulate plant model elements and will manage all requests and model-related queries from both humans and systems/modules within PEEE. Plant object-oriented model (POOM) is proposed to be the model repository within CAPE-ModE.

Using the proposed modeling methodology and environment, the design and implementation approaches of the target plant enterprise safety will be presented. Computer-aided plant enterprise safety management system, or simply CAPE-SAFE is presented as a complete solution to manage safety aspects within industrial enterprises. Such system comprises different modules to evaluate & follow-up hazards, synthesize operating procedures, manage safety regulations and legislations, manage safety data, and support and manage corporate safety training activities. The different modules can be managed and synchronized using a controller module.

To achieve the overall view of the automated plant enterprise environment, this book presents the information infrastructure of plant enterprise engineering environment (PEEE), which is intended to be a standard infrastructure to support lifecycle activities. The integration between the different modules of CAPE-SAFE with the corresponding engineering modules / components within PEEE will be presented. Information flows are showed to explain the utilization and operation mechanism of the target PEEE.

Towards the proper development of CAPE-SAFE, and in order to identify the actual user requirements from all stages within the plant lifecycle, a comprehensive system analysis practice has been carried out using object-oriented computer-aided software engineering (CASE) tool. The gathering and analysis of user requirements are essential to define the design specifications of the target CAPE-SAFE and to identify the integration points with other modules within PEEE. The proposed automated solution has been initiated as part of the collaboration with industry, a petrochemical plant where real user requirements have been gathered and analyzed while conducting the analysis phase.

In order to explain the mechanism and benefits of the proposed solution, case studies have been selected from HDS continuous plant and PVC batch plants. The related data of the selected plants are mapped into the structure of the different data groups within CAPE-SAFE modules.

The implementation of the proposed CAPE-SAFE in chemical/petrochemical plants or other manufacturing plants will reduce the operating cost and minimize the associated risk.

A prototype system framework of PEEE has been developed using Java and Visual Basic as object-oriented programming languages, while SQL server and MS-Access are used as RDBMS for data repository for the plant model elements. Such prototype solution will enable all plant users to utilize a standard user interface to manipulate plant lifecycle data/information/knowledge. The prototype system can be used as the framework to develop CAPE-SAFE as well as other components within PEEE.

1.2. Structure of the Book

In order to achieve the stated objectives, this book is organized into five parts:

Part I: Introduction

Part II: Plant Enterprise Engineering Environment (PEEE)

Part III: Computer-Aided Plant Enterprise Safety Management System (CAPE-SAFE)

Part IV: Utilization

Part V: Discussion, Conclusions, Recommendations, & Future Research Work

Part I: Introduction

Part I presents an overview of the book where scope, problem definition and solution significance are explained. It presents the background and similar research work done in this area. Also it describes the approach followed along with the theoretical and methodological framework developed to achieve the stated research objective. In this section, the plant safety model is presented as the base of designing CAPE-SAFE within PEEE.

Part II: Plant Enterprise Engineering Environment (PEEE)

In part II, the global picture of the plant enterprise engineering environment (PEEE) is presented with brief description about each component. More details about the modeling environment (CAPE-ModE) are explained in this section.

Part III: Computer-Aided Plant Enterprise Safety Management System (CAPE-SAFE)

Part III is the core of the book where complete design specifications of CAPE-SAFE are illustrated. This includes the analysis phase, the system architecture, integration with other components within PEEE, and the implementation of CAPE-SAFE.

Part IV: Utilization

In Part IV, the mechanism of the proposed solution will be presented where integration with other components within PEEE will be explained in more details. This includes integration with the modeling environment (CAPE-ModE), RCM-based maintenance management system, fault detection system, design environment, and with operation support system. Case studies are selected from continuous chemical plant i.e. HDS plant, batch chemical plant i.e. PVC, and upstream end of oil refinery process to illustrate the proposed solution.

Part V: Wrapping up

The closing section of this book presents discussion about the solution, conclusions gained from this study, recommendations, and the future work and enhancements to the current research work.

This book is partitioned into parts each part is composed of chapters. To have pictorial imagination of the organization of the chapters of this book, reader can use figure 1-1 to navigate through the different topics and research paths, which represents the line of though of the authors to address the different issues of enterprise safety management systems.

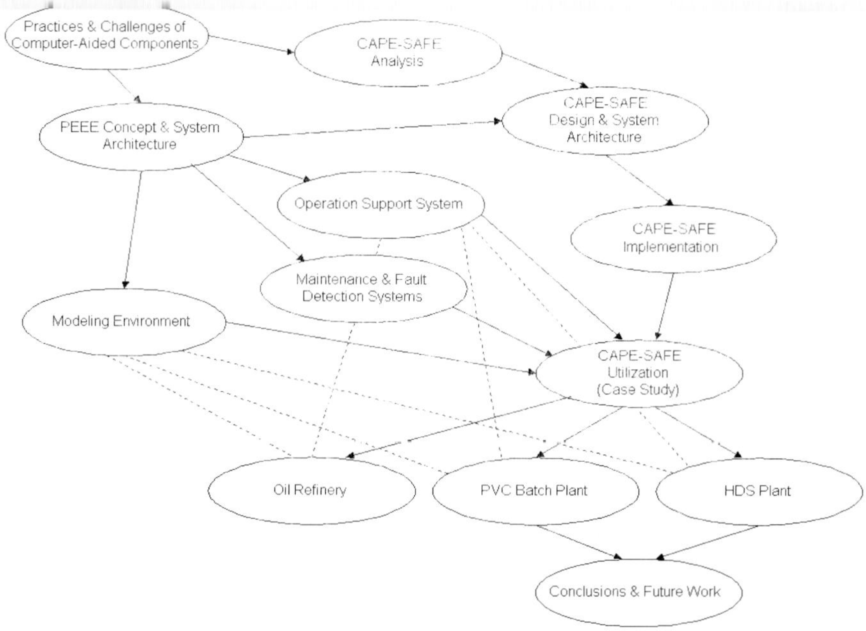

Figure 1-1: Research Paths

In the above diagram, the following research steps are required to achieve the book objectives:

- Current practices and challenges in computer-aided components focusing on those related to plant safety.
- Analysis phase using the limitations and challenges in the available safety software systems.
- PEEE the big picture.
- Design specifications of CAPE-SAFE.
- CAPE-SAFE Implementation.

- Plant modeling environment (CAPE-ModE).
- Utilization with Operation support system.
- Utilization with RCM-based maintenance.
- Case studies: PVC batch plant, HDS continuous plant, and Oil Refinery.
- Conclusions, recommendations, & future work.

There are different approaches to enhance process systems engineering practices. One approach is to enhance the system engineering methods used by formulating new methods or by modifying existing ones. The development and improvement of hazard evaluation methods i.e. HAZOP comes under this category. Another way to enhance process systems engineering is to improve the management and engineering of process knowledge either by providing more sources of knowledge (experts), knowledge representation, or through improving the learning techniques used to improve the underlying system/process e.g. neural networks. The third category that can be used to improve process systems engineering is the utilization of computer science and technology for example software tools like automated HAZOP. The combination of these three approaches will enable researchers and practitioners to achieve more valuable results.

In order to lead (and stay in a leading position) within the dynamic market competition, it is essential to improve the way the research is carried out. Allocating funds, and resources to carry out research projects can't guarantee the expected results and return on investment (ROI). Figure 1-2 shows our vision of researching the problem at hand.

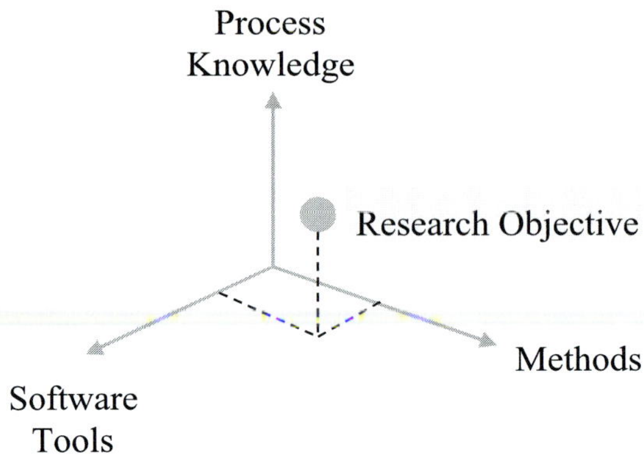

Figure 1-2: Research Objective

The philosophy of this research is to investigate and study the possibility of improving all the three dimensions and to design an efficient solution for process safety engineering.

1.3. Problem Statement

XYZ chemical/petrochemical plant is interested to manage all safety functions and activities throughout plant lifecycle via developing an automated safety management system. The proposed solution is required to be integrated with other modules/systems and tools available within plant enterprise engineering environment i.e. design, operation, management, and maintenance systems.

It is required to find a suitable way to achieve the target enterprise safety solution as integrated with other components within plant enterprise environment. Also it is required to define standards information & technology infrastructure to be adopted within the target plant enterprise engineering environment. In addition, the proposed solution should reflect top management goals and to adhere to the national and international health, safety and environment (HSE) regulations and legislations.

Data and knowledge models will be proposed to manage different model elements of plant enterprise safety. Knowledge engineering technique is required to enable the learning from past experience, with suitable integration with accident database. Plant personnel will use such safety-automated environment for safety-related training purposes using simulation practices, where they can analyze and understand real cases using real plant data. The proposed safety environment will also be used to support process design where it can be used for safety design and IPL design. The target safety environment will be used during the operation stage to support operator, where safety-related information and knowledge can be acquired and linked to the operational data and functions.

The proposed automated environment should include both qualitative and quantitative risk assessment methods, where decision engine can be used to decide the most suitable safety / risk assessment method at the different stages of plant lifecycle. In order to follow-up the risk assessment results a monitoring module should be used as a part of the proposed safety solution.

Towards the smooth implementation of the proposed plant enterprise safety management system, suitable knowledge modeling and representation approach is essential, which will enable plant personal to accumulate and interact with the plant model repository throughout plant lifecycle. A unified modeling language is required to represent the different model elements within the modeling environment so that it can be accessed and manipulated by other components and humans within plant enterprise engineering environment, including safety management environment. As per the current application technologies available now, a web-based systems is essential to enable the accessibility from remote or virtual nodes as connected to enterprise system network.

1.4. Safety Management

As with any management discipline Safety Management can be categorized into two areas:

1. Strategic
2. Tactical

1.4.1. Strategic

The Safety Management Strategy defines how safety and health is resourced, organized, measured, and monitored. The production of the Safety Policy and the Safety Plan is a key point to ensure that the strategics and procedures are implemented at all levels. High-level strategy must incorporate the implementation and regular review.

The Safety Management System will only work if it can proactively respond to the environment it strives to control. The planning and management of safety within any organization is key to its long-term viability.

The following issues come under the Safety Management Strategy:

- Safety Planning
- Policy Development
- Organizational Design
- Hazard Identification
- Competence Management
- Risk Management
- Benchmarking

- Standards Establishment
- Safety Culture Assessment
- Reliability, Availability And Maintainability
- Contingency Planning

1.4.2. Tactical

The challenge of keeping things moving safely on a day-to-day basis should be facilitated by the strategic plan but depends on the responsiveness of the organizations frontline.

Safety and Risk Management can assist the tactical aspects of business organizations to protect the workforce and maximize the operational performance.

The following issues can be viewed as part of the tactical side of safety management:

- Accident/Incident Review
- Crowd Control
- Emergency Planning
- Hazard Identification
- Fire Safety
- Contractor Management

The ideal safety management system will cover both aspects.

1.5. Benefits to Business

The proper development and application of enterprise safety management will return many benefits to business. The center for chemical process safety reported:

- Freedom to manage business (or "self-determination")
- Avoidance of major business loss
- Corporate responsibility, and
- Creating positive business values

Managing process safety requires a set of management systems directed at the prevention of major accidents involving hazardous materials. The focus of these management systems is:

- Technology: process safety information, process hazard analysis, operating procedures, safe work practices, and management of change
- Facilities: mechanical integrity and pre-startup safety review
- Personnel: employee participation, training, contractors, incident investigation, emergency planning and response, and auditing

As reported by researchers and industry that there are many motivating factors to invest in the development of robust and practical enterprise safety management systems. Among these, maintain company's existence, retains trust in clients, reduce operation risks and costs and major accidents, and lower the risk perceived by the investment community. Other benefits related to productivity such as increase the productivity and annual savings.

CCPS, ACC, API, OSHA, and EPA have developed management systems for process safety. In this book, fresh viewpoint will be discussed from the prospective of systems engineer.

1.6. Research Significance

This study is dedicated to design an integrated environment to manage plant enterprise safety activities and functions throughout the plant lifecycle so that it can ensure minimum risk and cost while operating plants. There are many attempts that are made to achieve such objective where safety assessment methods are automated and embedded or integrated with enterprise management systems. Most of these attempts established integration between safety components and process design, modeling and simulation components. There are critical issues that are highlighted from such conventional automated solutions, such as lack of systematic integration between process / micro level and the enterprise / macro level.

The proposed modeling methodology enabled such integrated view between micro and macro levels, which can be extended to nano level as well. The study of enterprise safety using integrated hierarchical model will enable the proper analysis of safety activities as inherited from one model element to another, which will also reflect aspects from the inherent safety design.

The proposed enterprise information infrastructure enabled the smooth integration among the different efforts made to automate one or more safety method or function / activity. For example, automated HAZOP and FTA / ETA can be redesigned so that it can be simply integrated within the proposed enterprise safety management system using the proposed standard engineering middleware and standard and structured hierarchical model repository. In addition, the integrated enterprise solution will enable realizing real time response and support the management of change.

The expected or intended integration doesn't mean interfacing technologies within the information technological infrastructure or integrating systems within the information systems infrastructure within the plant enterprise. But rather both aspects are addressed simultaneously, in addition to the analytical part, which can be done only through proper system analysis on the basis of robust modeling methodology, as part of the software engineering lifecycle. The integrated requirements from different views such as designer, operator, manager, and maintenance engineer, etc. along with the requirement analysis approach helped achieving efficient enterprise safety management system.

Another significance of the proposed study is that it has been carried out as part of the collaboration with industry where it has been revealed by senior safety engineer of petrochemical company, during several interviews, that the highest priority and concern of the company's top-management is the design and implementation of practical and successful enterprise safety management system that to be integrated and incorporated within the underlying plant enterprise where it covers all safety functions and activities throughout the plant lifecycle.

This book answers the management's top-concern and propose integrated safety environment as part of enterprise engineering environment PEEE. The results of this book can be used efficiently by both research and industry to analyze enterprise safety for better enterprise safety management systems and to realize automated safety functions as integrated with other systems and components within the plant enterprise engineering environment.

It is essential to answer the question: why we need automated safety functions? As reported by Srinivasan that the Occupational Safety and Health Administration's (OSHA) process safety management (PSM) standard requires initial PHA of all the processes covered by the standard to be conducted. By one estimate, this requires PHA of about 25,000 plant sites to be completed in the United States at an estimated cost of two billion dollars (Srinivasan et al., 1998). So by automating safety functions it is most likely to reduce the time and cost required performing such safety functions. Similarly, by looking to the resources allocated to maintain, document, apply, and assess safety regulations in plant sites, one can understand that these activities are similar among plants and repeated in different stages within the lifecycle of the same plant. By automating such activities and enabling the sharing of knowledge and information among plants, it will reduce the operating cost and improve the level of accuracy of safety data. Safety training is another opportunity of reducing the cost and improving the level of plant personnel by offering continuous safety training using real examples from the current and/or similar plants as integrated with operation, design, and simulation environments. By offering integrated computer-based safety training, the training cost can be reduced to minimum, while ensuring more efficient training using real plant data and scenarios.

The significance of this book can be highlighted by answering the following three major questions:

1. What is the optimum/best design of the system architecture for the integrated plant enterprise environment? This has been answered by proposing PEEE.
2. How can we achieve the optimum/best design of the integrated system that manages plant enterprise safety throughout its lifecycle? The answer is CAPE-SAFE.
3. How to integrate the proposed system within the plant enterprise integrated system, so that it can share data and to be beneficial to all humans and components within the plant enterprise environment? The answer is the specification of the integration between CAPE-SAFE within PEEE.

In particular, this study provides significant findings in the following areas:

- Object-oriented modeling of plant process where plant model is represented in object-oriented manner. This allows the representation of plant operation, behavior, as well as accumulating other plant lifecycle activities, such as safety, within central plant object-oriented model.
- Propose conceptual framework to represent safety aspects in plant safety model where we could manage to abstract plant lifecycle safety aspects and propose opportunities to represent within plant enterprise engineering environment.
- Design a complete picture of plant enterprise engineering environment to accommodate for engineering and non-engineering sides.
- Design framework of the complete solution for plant enterprise safety management system that can manage the plant lifecycle safety functions and activities.
- Design a new method of representing the fault propagation method, which can be used to represent the plant fault propagation model within the plant OO model, and to automate the process of hazard evaluation using computer program. This method can be used effectively to automatically convert all existing HAZOP results into electronic and structured format. This will enable plant design and operation systems to utilize hazard evaluation results to assist designer and operator to perform their functions efficiently.

The validation of these questions and proposed ideas will be verified using the case studies selected from chemical / petrochemical plants to show the mechanism of the proposed integrated automated solution and to highlight its benefits.

2. Background

In order to start our research project from the point where other researchers have ended, it is essential to review the work done in the area of integrated solutions for plant automation as well as in the area of computer-aided safety tools. This will help us learning from the latest experiments and capture their features and advantages and solve or avoid their drawbacks. Also it is essential to review the work done in the area of plant modeling from all aspects as they are related to plant safety engineering solutions.

2.1. Industrial Practices

Currently, most of the chemical/petrochemical plants are organized into departments and sections, where each department is mainly managing one or more business function. The difference between managing business function and carrying out the business function is that one department may manage one functional area, but it requires other departments to carry out some tasks. For example, maintenance department manages the plant maintenance, however, operator; from operation department; carries out some of the maintenance functions while operating the plant. To compete in the market competition, almost all owners of the plants decided to modernize their business by introducing information systems and technologies i.e. automated systems, and by simplifying the business processes within each area. We can see the majority of the plants are implementing software packages that cover one or more of their business functional areas i.e. ERP. There are many limitations that face this direction, for example the integration among these packages requires tremendous efforts, while the advancement of the current technology represents another challenge. The difficulty of integrating the engineering and non-engineering views has created a gap between these two domains and hence limiting the designer from getting the benefits of using the procurement data while performing the design rationalcs. This also disturbed the buyers from getting clear and correct design specifications to purchase the right equipment for the right operation/process. From the other side, it is essential to structure the plant lifecycle data to enable smooth data exchange among the different plants, which will help plants to use useful data from other plants. STEP has made remarkable efforts in this regard so that it becomes easier to exchange plant data among different systems and domains i.e. between design, maintenance, and operation systems. Internet-based solutions are another trend that is currently adopted by most of the industries to overcome some of the integration limitations where middleware i.e. CORBA has been used as a backbone to integrate the different heterogeneous systems.

From safety engineering point of view, almost all plants are decided to adopt and implement safety management system, which is managed by safety division or department within the enterprise. Safety training is one of the most important issues that require all plant personnel to be trained on the latest safety practices and technologies. The initial estimate for one organization to organize continuous safety training courses is thousands of dollars per year. This mobilized many companies to utilize computer-based training (CBT) and web-based training (WBT) approaches. Safety regulations and standards are another area of interest where many plants started writing their own standards in view of the national and international safety regulations (i.e. OSHA). Safety assessment is another area of interest where automated tools are required to carry out and assist in performing the hazard evaluation practices. In this automated HAZOP is used in few cases to carry out the hazard evaluation technique.

2.2. Literature Review

This research work proposes automated safety environment as integrated within the plant enterprise engineering environment. This requires integration among the automated design, operation, and safety environments as integrated with the plant lifecycle modeling environment.

Within the area of plant lifecycle modeling environment, Tokyo Institute of Technology has developed a multidimensional modeling on the basis of object-oriented approach, called MDOOM (Naka et al., 1995). In this approach plant is modeled in three dimensions: behavior, physical, and operation. A design modeling environment using G2 has been developed using such approach (Batras, 1999).

Another plant modeling approach is proposed by Pohjola, which describe plant as group of interrelated objects. Each object is described as: Purpose, Structure, States, and Performance. This method is referred in the literature as PSSP modeling methodology. This method has been employed to develop computer-aided tools for design and waste management.

For the purpose of the design of engineering systems, another approach is proposed by Gofuku (1994), which is based on goals, functions, structure, and derivation of behaviors. This approach is implemented using Smalltalk-80. Such approach combines Multilevel Flow Modeling (MFM) and the Hybrid Phenomena Theory (HPT).

In the area of plant enterprise integrated solutions, although studying the complete picture of plant enterprise is considered as wide area of research as it is related to different disciplines and areas of research, but till to date there are many attempts that are trying to define an optimum and efficient integrated solution for plant integrated systems.

In the field of plant operation, object technology is employed to develop a computer-integrated environment for process operation (Qian et al., 2000).

In the area of computer-aided safety engineering tools, there are many attempts that have been done to automate one or more safety function. One major safety function is hazard evaluation. HAZOP, which is hazard and operability study, has been developed as a systematic approach to identify all possible deviations from the way the design is expected or intended to operate the plant and all hazards associated with these deviations. Automated HAZOP is one major area of research where many attempts have been made to develop computer-aided tool to automate the hazard evaluation process using HAZOP method. This has been varied between fully automated tools, which is designed to generate the HAZOP complete results, while there are other tools which is used to along with the HAZOP expert team to register the results, findings, and to provide useful information to reduce the time and efforts required to run normal HAZOP process. HAZOPEX system highlighted the roles of computer and human analysis in performing computer-aided HAZOP study (Karvonen et al., 1990). FAULTFINDER computer program has been developed by Loughborough University based on modeling of fault propagation, and fault tree synbook (Kelly and Lees 1986a, b, c, and d).

2.3. Commercial Products for Computer-Aided Safety Engineering

There are many products available now in the market that can manage different activities of the plant enterprise safety, for example CARA (Sydvest), which is a computer-aided fault tree analysis tool for quantitative hazard evaluation, and PDS Tool for Reliability Analysis of Safety Functions, both are categorized as reliability, availability, maintainability, and safety engineering computer-aided tools. Also Kyrass (Sydvest) is used for Identify and Follow-up Hazards and Weaknesses in Enterprise. Anther integrated computer-aided safety management tool called SafetyManager (Ergosystems) has been proposed with integrated modules for safety audit, safety training, risk assessment, safety reporting, and safety legislations and regulations. The other product is CDM Range (Ergosystems), which includes Health & Safety data files, planning, and wider scope for process designers. The key success factor for these computer-aided tools to work is the integration with design, operation, and simulation environment. Still most of these commercialized safety tools are far beyond the industry expectations and real needs.

3. Theoretical & Methodological Framework

3.1. Research Approach

The ultimate target of this study is to develop an efficient and complete safety management system that manages plant safety throughout the plant lifecycle as integrated with plant enterprise engineering environment. In order to achieve the research goal as well as the earlier-stated objectives, it is essential to understand and study the concepts of modeling and plant enterprise engineering environment using object-oriented approach, and then employ these concepts to design plant enterprise safety management system. This book can be viewed as a hierarchical structure with major research activities that need to be carried out, as shown in figure 3-1.

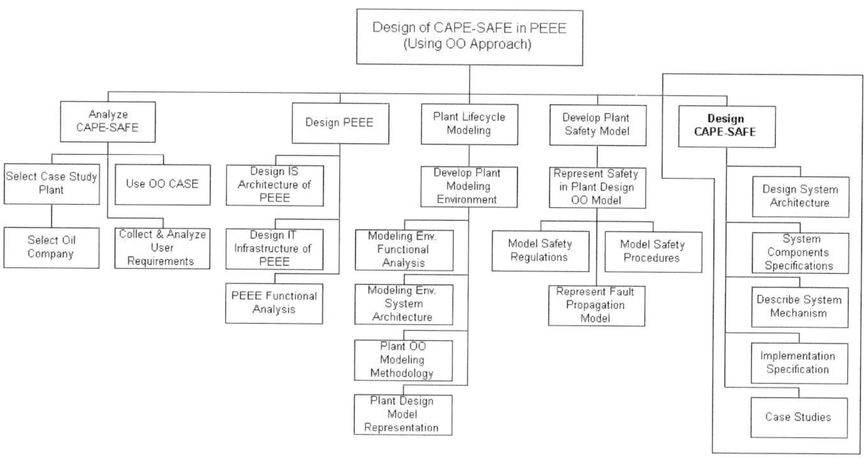

Figure 3-1: Research Approach

In this figure, and in order to design CAPE-SAFE, it is essential to define the approach to analyze user requirements, design the global picture of PEEE, define plant lifecycle modeling approach including OO modeling methodology, design framework for plant safety model to represent safety aspects in plant lifecycle, and to represent fault propagation model in an efficient way that supports the automation of hazard evaluation and safety model representation within the proposed OO modeling environment. Finally, design a framework to realize CAPE-SAFE in PEEE.

In this section, highlights on plant lifecycle modeling using object-oriented approach as well as the conceptual safety model that covers all safety aspects within the plant lifecycle will be explained.

3.2. Object-Oriented Modeling Framework

Model is a formal description of a system or a process. Entity 'M' is a model of a system 'S', if 'M' can be used to answer questions related to characteristics of the system 'S' (Woods, 1993). The model development is intended to represent one or more points of view for the underline system (or plant) (Gabbar, 2000g). This means developing a model to represent a system can't be achieved unless we specify which point of view(s) such model represent the system. Information model of a system represents the information organization point of view of a given system. Other examples of models are: fault propagation model, data model, knowledge model, etc.

The modeling methodology is based on abstracting the plant objects in three views: static, dynamic, function. This can be represented in diagrams associated with the different model elements.

3.2.1. OO Model

In the real world, objects can be viewed in three dimensions: static, dynamic, and function. Table 3-1 explains the meaning of each of these dimensions (Gabbar, 2000g):

Table 3-1: Object-Oriented Model Views

Static	To explain what is there? And it comes under two factors: structure and topology. Structure represents the subcomponents of the object, while topology represents the interconnection among the different objects.
Dynamic	Or behavior, to explain what happened and when it happened?
Function	Or operational, to show the algorithm or the computations of how it happened

3.2.2. Object-Oriented Modeling Using UML

In order to construct the model corresponding to a system or process, it is essential to identify the three dimensions in a formal, systematic, and standard manner. Moreover, OO CASE tool can be used as a modeling framework and as a repository to store model elements. In order to construct the OO model a methodology is required to systematically construct the underline model. Unified modeling language (UML) has been selected as the standard (informal) language to represent the model. In such model there are four levels of abstraction in the metamodel architecture: meta-metamodel (the infrastructure for the metamodeling architecture. Defines the language for specifying metamodels), metamodel (An instance of a meta-metamodel. Defines the language for specifying a model), model (An instance of a metamodel. Defines the language for specifying an information domain), and user objects (An instance of a model. Defines specific information domain). Figure 3-2 shows the modeling framework using UML.

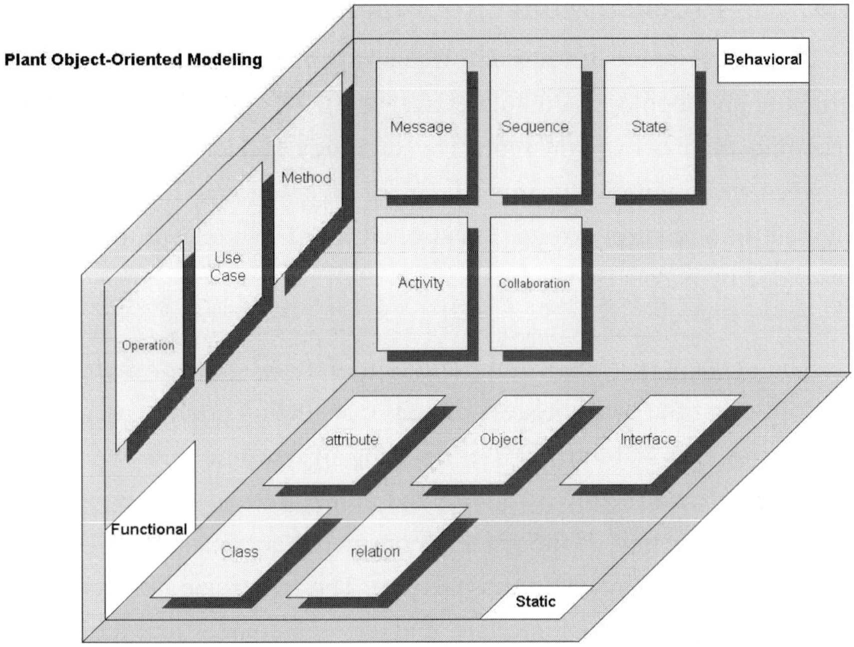

Figure 3-2: Plant Object-Oriented Modeling Framework Using UML

3.3. Plant Lifecycle OO Model Representation

The main purpose of developing plant lifecycle model is to understand and analyze the different activities within the plant lifecycle and also to develop the different systems to automate the different functions within the plant lifecycle. As the plant lifecycle is complex so it is essential to have a simple approach to develop such complex model. Our proposal in this book is to use the simple OO modeling concept to construct the plant lifecycle models. By representing the plant model in UML-based repository will enable converting P&ID into active plant object-oriented model, which is informative about all attributes and specifications of each model element i.e. pump, reactor, etc.

3.3.1. Process Model Representation

A process plant has a static dimension that can be represented as structure and topology (Pohjola et al., 1997). The structure expresses the decomposition of a plant unit. The topology reflects the connectivity between units within the same level of abstraction. The plant can be divided into control group units (CGUs). The concept of CGU is described by Naka (1999).

The concept of connection point is important to allow more precise description of the plant process design. Connection point expresses both merging of two or more lines or splitting of one line into two or more lines. The plant structure is represented in hierarchical manner where the top-level abstraction of the plant is considered as a master CGU, which is represented within UML as a package. The inputs and outputs of fluid flows are represented as interface classes with sources and destinations connected with the main CGU via fluid flows. P&ID is represented as class diagram, where fluid flow is represented as fluid flow associated with stereotype of the fluid. This will link the fluid to the fluid class, which includes all attributes. Each fluid flow in any level is associated with the following information: (1) "Signal", which is a specification of a asynchronous stimulus between objects and is processed by the StateMachines, (2) "Destination element", which indicates the destination of the fluid flow, and (3) "Metaclass", which links the fluid flow to a given fluid class (defined within the metamodel level).

3.3.2. Plant Operation Model Representation

The operation of each equipment will be represented as an associated method with the class that represents this equipment. This will enable the operational models to be associated with the static model of the plant. This means that the P&ID can be represented along with the operational models of the different equipments, which supports the concept of operational design approach (Naka, 1995). This can be useful for the operating procedures synthesis practices to be automatically generated form the plant OO model.

3.3.3. Plant Behavior Model Representation

The behavior of the plant process can be represented as states and transition among the different states. Also messages and collaboration among the different classes/objects can represent behavioral aspects of the plant. Equations controlling the transition from one state to another can be represented using object constraint language (OCL) as an associated code with the different classes and states.

3.4. Plant Safety Model

There are a variety of computer programs that claim its superiority in safety engineering. The key of evaluating such solutions is how plant enterprise safety model has been abstracted and how the solution is fully integrated with other components within the plant enterprise engineering environment. In this section we are proposing a conceptual plant safety model that covers the general concept of plant safety throughout the plant lifecycle.

Managing the plant lifecycle activities requires an automated, integrated, and concurrent environment that supports information sharing and enable the integration among the multiple technologies. Such environment augments both technical engineering systems as well as enterprise commercial (non-engineering) systems. This can simply describe the proposed Plant Enterprise Engineering Environment (PEEE), which is intended to manage all functions and activities of the plant by sharing and reusing the resources and knowledge. Modeling is the most efficient approach to achieve such environment. Constructing the plant lifecycle central model requires analyzing the different views that exist throughout the plant lifecycle via constructing a plant lifecycle object-oriented model. The object-oriented approach is an effective approach to construct such model. The plant-wide object oriented model includes all plant objects associated with multiple views to express its static, dynamic and functional syntax and semantics (Gabbar, 2000a,b). The developed model can be used interactively by both humans and systems in the target environment (PEEE). For example, within the plant design environment, process design model is maintained during the design stage, which is linked to the operational model, this will help the process designer to effectively complete the process design. The plant lifecycle modeling can also support the concept of operational design (Naka, 1995). From the other side, operational systems (i.e. for operator) can acquire process design model (plant structure, behavior, and/or operation) from the central model while operating the plant. This can be used for process shutdown, malfunction, or any abnormal situations. Hence, building the central model is considered as the cornerstone to achieve PEEE. In such environment safety is considered as a common objective within all activities throughout the plant lifecycle. In order to ensure plant safety it is essential to consider process safety throughout the plant lifecycle. The ideal way to achieve that is to consider the plant as an entity that is composed of objects, which are modeled using object-oriented modeling approach where each plant object is expressed as static, dynamic, and functional views. By defining the safety aspects for each model element, the plant safety model can be realized. In this section the approach of representing plant safety model will be explained as integrated within the plant lifecycle model using object-oriented approach within the PEEE.

3.4.1. Plant Safety Modeling Approach

Safety engineering is based upon fundamental principles and rules of practice, which involves the identification, evaluation, and control of hazards in man-machine systems (products, machines, equipment, or facilities) that contain a potential to cause injury to people or damage to property (Nelson & Associates). One feature of the safety engineering is modeling or constructing the plant safety model. In order to construct the plant safety model, it is essential to identify all safety aspects within the plant lifecycle. By identifying the plant safety aspects within the plant lifecycle model, the plant safety model can be achieved. Safety aspects include risk assessment considerations, which are raised by auditors within the plant enterprise to ensure compliance with enterprise procedures (or standards). As the safety is a common and essential objective for all activities, our approach is to find a way to represent safety within each model element of the plant-wide model. Safety aspects include safety procedures, which are mapped to the allowed-operations (or not-allowed-operations) for each plant object. Also, safety aspects include safety features or characteristics, which are associated with each plant object. The whole dilemma is to develop a systematic way to represent safety aspects and to realize them starting from the plant design, and to ensure its fulfillment throughout the plant lifecycle (i.e. procurement, construction, operation, and disposal). It is dilemma because these activities are implemented by different parties via different systems in different environment. From the other side, accidents and incidents and other historical safety-related data are essential to develop the scenario-based model that can be used for simulation-based practices. Safety equipment and devices are represented as the control systems utilized within the process, where it is designed, constructed and then operated similar to the process equipment itself. In this case (i.e. for safety equipment and devices) it has also safety considerations that to be managed throughout the plant lifecycle. The incremental object-oriented plant safety modeling will enable the addition of views and model elements throughout the plant lifecycle, as in figure 3-3.

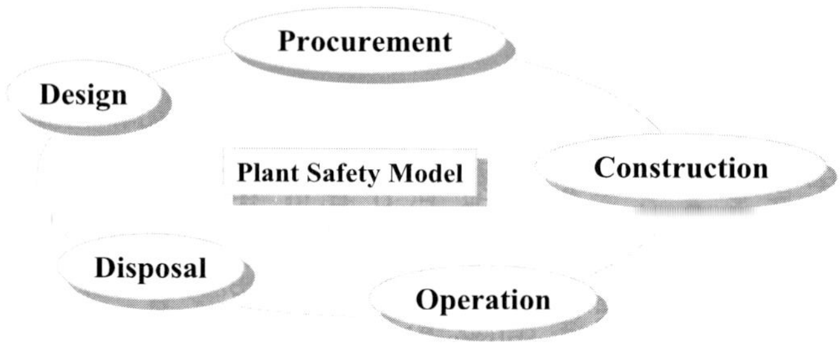

Figure 3-3: Plant Safety Model within Plant Lifecycle

3.4.2. Plant Safety Framework

Safety objects are developed within the process to keep the process in safe condition and to avoid hazards. Safety objects are mapped to model elements such as operating procedures, alarms, sensors, barriers, etc. It is essential to elaborate more on keywords and terms used hereinafter. **Event** is a dynamic change in the state that occurs to a system element (to an object within the plant domain). The event can be defined also as an occurrence related to equipment performance or human action, or an occurrence external to the system that causes system upset (CCPS, 1993). If the change is matching (consistent) with the intended design so it is normal event, otherwise, it is abnormal event. Hence, **abnormal situation** is that situation which is not matching with the intended design i.e. expectations. This also defined as **fault event**, which is a **failure situation** resulting from logical interaction. **Hazard** is an inherent chemical or physical characteristic that has the potential for causing harm or damage to people, property, or the environment (CCPS, 1993). Hazard is directly related to forms of energy, since system component or personnel damage or injury cannot occur without the presence of some hazardous energy. **Accident** is an unplanned dynamic (typically multi-causal) event or sequence of events that results in undesirable consequences (CCPS, 1993). It is an incident with specific safety **consequences** or impact. The transformation from cause to consequence can be viewed in figure 3-4. Failure is a stoppage or malfunction of a component, described as an error in the operation associated with a physical object within the plant domain. Failure can be represented within the plant model as a state associated with the physical object (e.g. **VALVE.STATE.OPEN**), while fault is an operation error (or deviation) in one of the plant components or parts. Fault is the erroneous phenomenon, which describes the problem. Logically, fault is due to cause of failure in another plant component or part. By analyzing the process component failures, any process or part of a process has four major symptoms (of operation mode): normal, failure, fault, or hazard. Operation mode may result from any combination from the four different possibilities: environmental effects, design errors, operation (operator) errors, or maintenance errors (as shown in figure 3-5.

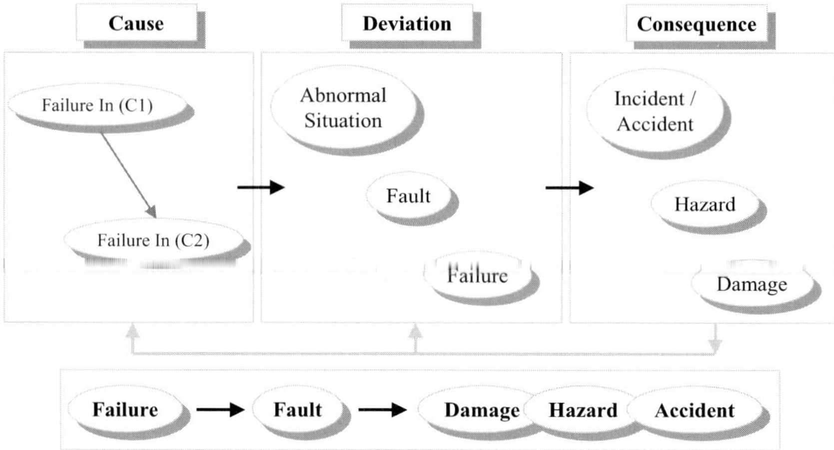

Figure 3-4: Cause-Consequence Relationship

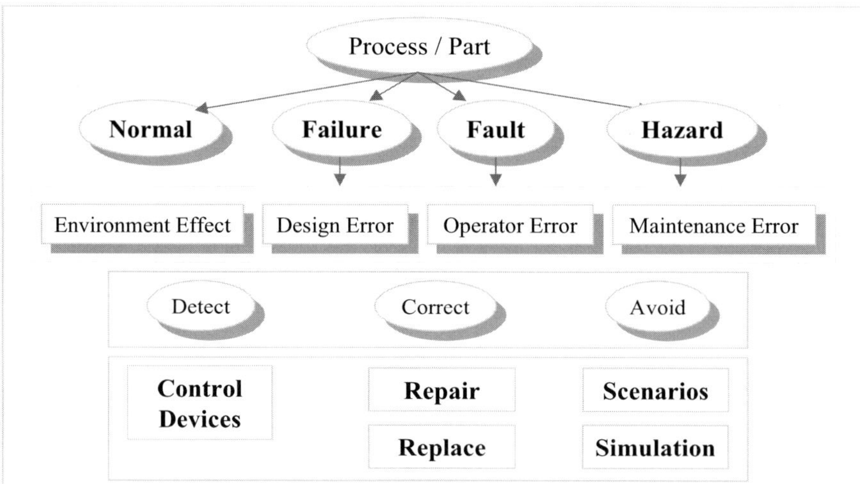

Figure 3-5: Process Safety Failure Hierarchical Reasoning

3.4.3. Plant Safety Model Components

Plant safety model can be represented as collaborative building blocks, as in figure 3-6, that can be integrated around the main functional areas within the plant enterprise. *First* is the "**safety procedures**" where all safety regulations and jobs are explained and attached to plant object, action, or state transition. *Second* is "**safety specifications**", which describes the plant objects characteristics and attributes. *Third* is "**safety historical data**", which capture historical data in standard format that is readable to plant systems. *Fourth* is "**safety common data**", which includes all common data that can be shared and exchanged among different plants and components/systems. For example, failure codes can be shared among plant safety systems as standard failure hierarchical codes. *Fifth* is the "**safety scenarios**", which is an intelligent component to envision the safety level by building different scenarios with its safety level ranking. *Sixth* is the "**safety devices**", which is part of the process control system to monitor the process variables and to detect and recover any abnormal situation.

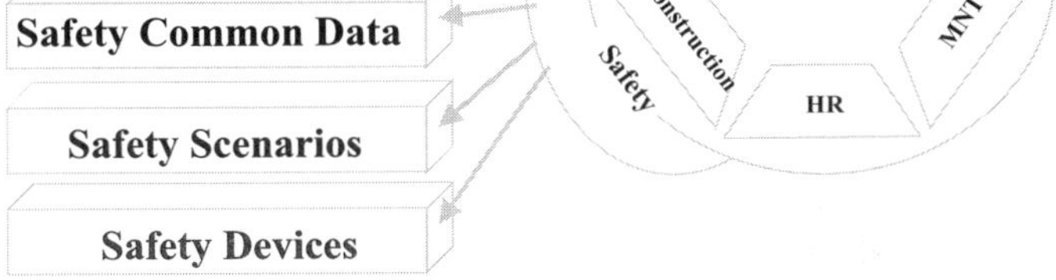

Figure 3-6: Plant Safety Building Blocks

As the proposed plant safety model is aiming to cover all activities in the plant lifecycle to ensure safe plant activities, it is difficult to achieve that without having a systematic approach to abstract plant lifecycle activities into object-oriented model with safety aspects and considerations embedded. It is possible to incorporate the process safety management components into the plant-wide model using the object-oriented approach, for example safety procedures are mapped to actions associated with transitions between object states. In summary, by realizing the building blocks of the plant safety model onto the five stages of the plant lifecycle, safe plant lifecycle can be achieved.

3.4.4. Plant Safety Management System within PEEE

Figure 3-7 shows the proposed plant safety system architecture within PEEE. Plant central object-oriented model contains all plant objects that are managed via the plant model engine. The main responsibility of such engine is to maintain the integrity and completeness of the model. Plant physical objects are stored in different forms i.e. database, knowledgebase, or data files and can be accessed via the standard middleware engine. Plant safety agents are interacting with each other and with other components in the proposed architecture to perform the plant safety functions. Application level integration supercedes the data level integration. This will enable the use of the object technology to build standard objects and components that handle services, which can be common to all applications. For example, *create-equipment* service can be used in all applications (process designer, operational systems, and maintenance and procurement systems). Hence writing such service using the middleware technology with object-oriented embedded code can enhance the productivity and interoperability of the developed components.

The proposed plant safety component architecture can be viewed as in figure 3-8. In such figure, equipment data are centrally maintained using the maintenance agent, while process design model (including P&IDs) are managed via process designer agent. The backbone of the developed solution is mainly a middleware engine, which can reduce the gab among different technologies and components while managing the communication among them. Common data and historical data agents are responsible for defining the required data and managing them throughout the plant lifecycle. These two agents communicate with the operational, design, and maintenance components to debate the scope, validation and lifetime of the stored data elements. Process simulation engine maintains the simulation model based on the process design and process safety considerations. Maintenance agent manages the maintenance plans and jobs with respect to the process safety considerations and plant operational condition.

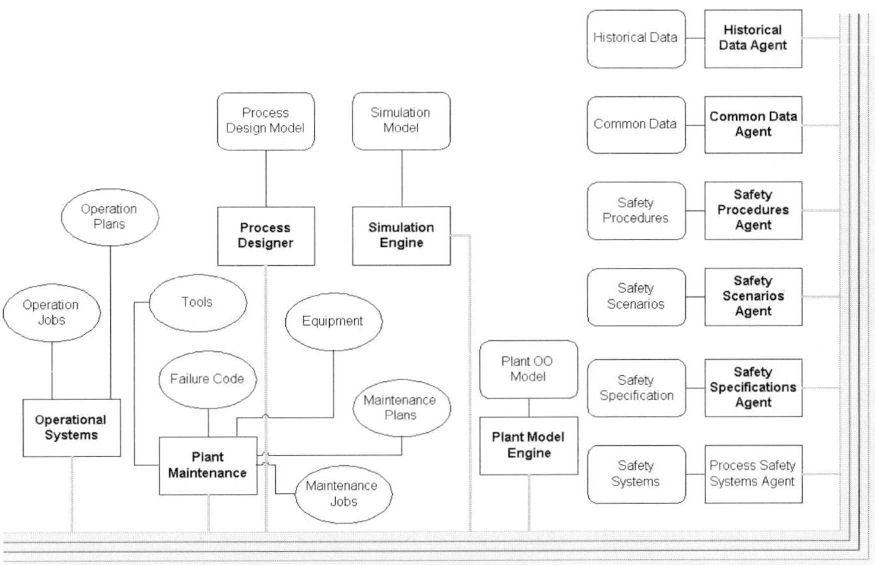

Figure 3-7: Plant Safety System Structure

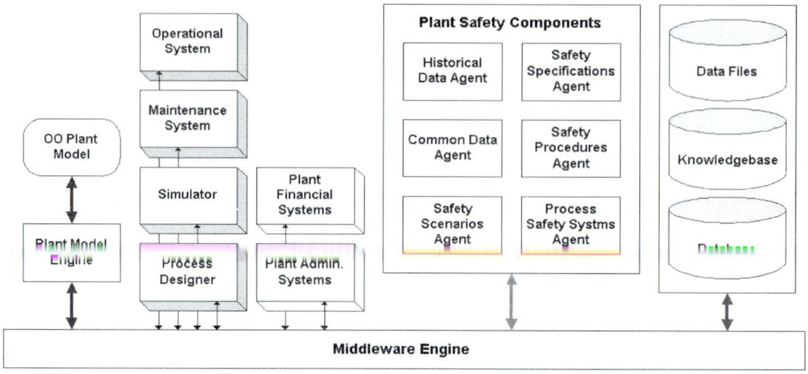

Figure 3-8: Plant Safety System Architecture

3.4.5. Plant Safety Procedures Component

Safety procedures are set of instructions to be followed in certain conditions within the plant lifecycle to keep plant safe (Van, 1999). For more details, let us analyze a sample safety procedure from OSHA (OSHA Publication No. 3132) (see Ref. OSHA Administration) as follows:

> **Process Safety Information:** *Employers must complete a compilation of a written process safety information before conducting any process hazard analysis required by the standard*

The abstract of the above safety procedure could be as: "**Safety Procedure**" is associated with **Hazard Analysis** Activity as a **precondition**. **Employer** is an *actor* associated with "**Process Safety Information**". This is mapped to the object-oriented modeling approach by using Use Case and Collaboration diagrams. The completing the process safety information prior to conducting Hazard Analysis is implemented as a constraint associated with the "Start Hazard Analysis" use case, as in figure 3-9. Use Case diagram is used to show elements from the use case model. The use case model represents functionality of a system or a class as manifested to external interactors with the system, while collaboration diagram may be attached to an operation or a use case to describe the context in which their behavior occurs. The actual behavior may be specified in interactions, such as sequence diagrams or collaboration diagrams (the different modeling diagrams are explained in details in UML reference). A collaboration diagram may also be attached to a class to define the class's static structure. In general, safety procedures may be represented as *constraints* associated with plant object. Constraint is an extension of the static model of the plant to restrict possible states i.e. transition between states. Safety procedures constraints can be represented using OCL within the object-oriented unified modeling language (UML). Despite the reported drawbacks about OCL, it can be used to specify the metalevel as well as the user level specifications of the constraints associated with model elements. This approach facilitates the automatic conversion from the safety procedures into the formal plant constraints, and hence enables the automation of the safety procedures from the plant model. From the other side, safety procedures could be represented as set of recommended actions associated with states of the plant object, for example how to disconnect a process safely to avoid any explosion, toxicity or flammability, can be represented as an associated OCL with the specified process. OCL can be included in the "ToDo Errors", "ToDo Warning", "ToDo Unused", Comments, or as Description. From one stage to another within the plant lifecycle, safety procedures can be represented, incrementally, using the above mechanism. By this way, it is likely to have at any point of time, all safety procedures in place in the plant-wide model. The plant operating procedures are commonly grouped as per the following group: Commissioning, Start-up, Normal shutdown, Emergency shutdown, Preparation for maintenance, maintenance, Sampling/tapping/draining, and Inspection/monitoring/checking. Where the unit of the operating procedure is composed of: Purpose, Action, Time, and Information. Example of the safety operating procedures is as follows: "Before starting the ammonia loading, all the manual trips will be checked with the A2 supervisor."

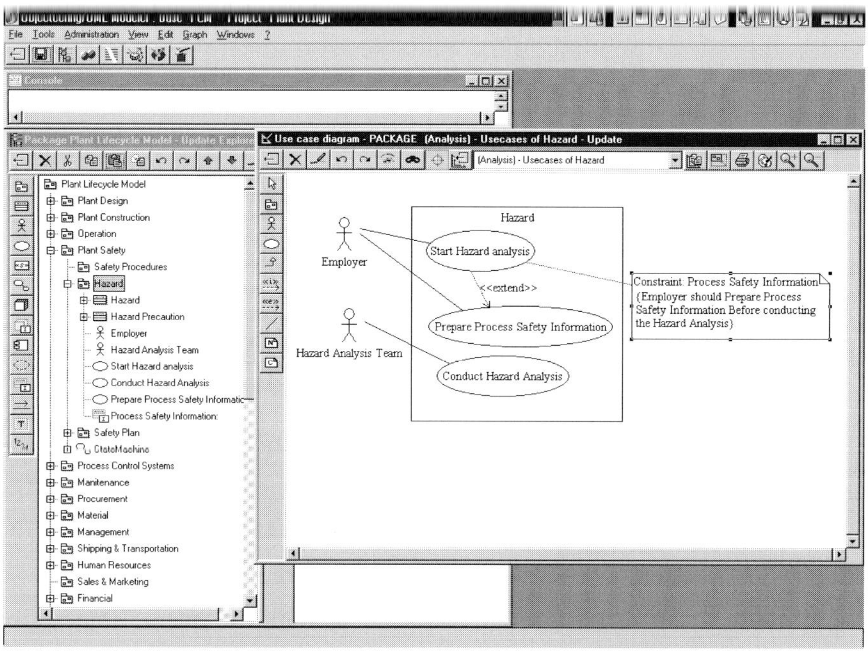

Figure 3-9: Use Case Diagram Representing Hazard Analysis Safety Procedure

3.4.6. Safety Specifications Component

Safety specifications are characteristics associated with plant objects (normally physical objects) to describe features needed for safety considerations. In fact, specifications describing physical objects can be considered as part of the safety specifications since any deviation from these specifications may lead to abnormal situation. Accordingly, safety specifications are considered as attributes describing the physical object. By reviewing the process data sheet, it is possible to represent the safety specifications as attributes associated with instances of the physical objects (classes) to be able to specify the required attributes for different states of the objects. For example equipment class in the first level of the metamodeling can be used as a stereotype to specify all equipment classes like pumps. Then pumps will be instantiated to have pump1 and pump2 with the inherited safety specifications as associated attributes. This has been implemented in the object-oriented model as in Figure 3-10. As an illustrative example, the "temperature_inside_pump" is an attribute associated with the pump class and is inherited to all pump objects. As a part of the safety specifications of the pump, the temperature of the pump should not exceed "temperature_upper_limit", this can be implemented by adding an attribute to the pump called "temperature_inside_pump" and associate a constraint as ("temperature_inside_pump" < "temperature_upper_limit").

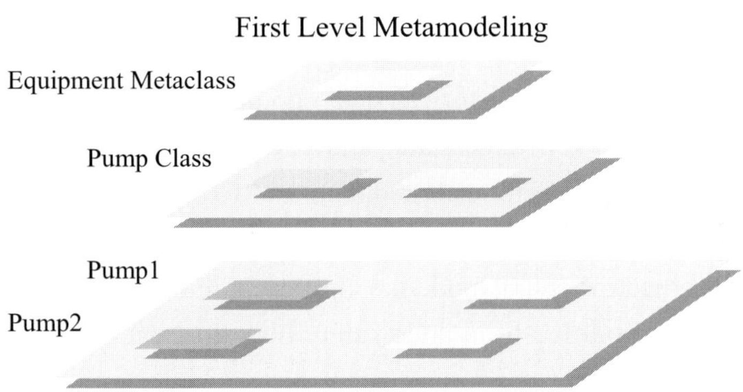

Figure 3-10: Process Equipment Recursive Metamodeling

3.4.7. Safety Historical Data Component

The idea behind expert systems is the ability to learn from others experience. Successful safety systems are those systems, which can learn from others (with accepted response time) with intelligent behavior. Most of the real time applications are linked to historical data sources (i.e. databanks), which contain historical and summary data (useful information from previous experience), to improve their efficiency.

Historical data is previously processed data and/or data related to actions/activities from the past. The data warehousing is used to capture history data (static and dynamic data) and store them in different multiple views to be ready for analysis (called data cubes). Such data cubes are designed based on the user views where user (superuser or analyst) navigates (drilldown) the data cubes based on the required analytical results. By including plant historical data group in the plant safety model the process safety can be improved. For that, it is proposed to start the historical data component with the base safety data entities and incrementally accumulate along with the plant lifecycle. The safety historical data component includes mainly two parts: incidents/accidents part and plant safety related data. By analyzing the historical data cubes, useful relationships among plant objects can be extracted with respect to process safety, which enhance the process safety and speedup the learning curve of plant safety systems to avoid hazards, failures, or abnormal situations. To ensure smooth data exchange among plants and heterogeneous systems, historical data should bc in a standard format, where all systems can understand and exchange safety historical data. STEP and PISTEP offers such standard and exchangeable formats that can be utilized to implement such component (NIST, ISO, and PISTEP). From the object-oriented modeling approach point of view, the plant model can be modified to include safety historical data in two different ways: state diagram, where specifying history state (either shallow or deep history state), or as dependent class associated with plant class. Accident/incident historical data can be used to develop scenarios while conducting the hazard analysis/assessment, which enhances the process safety by considering the previous events (cause and consequence with respect to plant type, failure type, etc.). Figure 3-11 shows the first level architecture for the historical data model, which is used within the plant safety framework. Safety historical data are refreshed incrementally from the online plant systems using predefined scripts with predefined criteria from different levels from the plant model. Historical safety data are exchanged among remote plants using standard data exchange format (DEF), which requires plant models to be standard on the basis of UML.

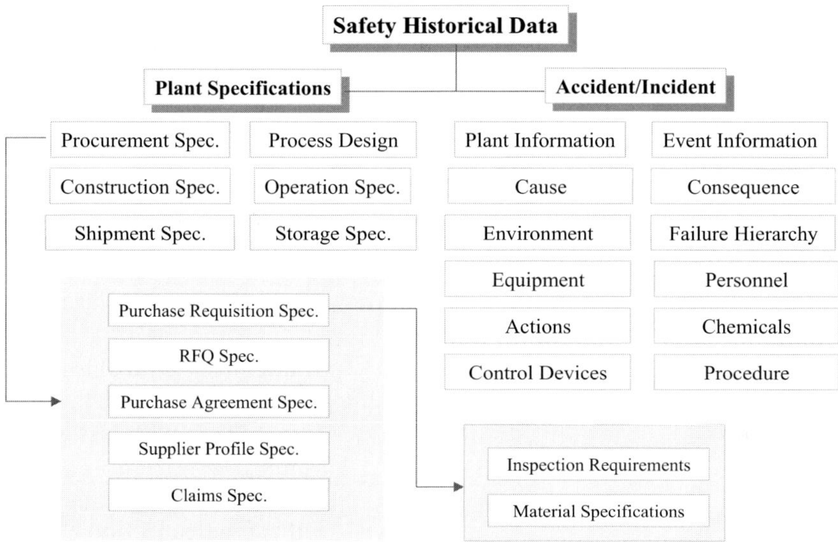

Figure 3-11: Safety Historical Data Modeling

3.4.8. Safety Common Data

As a part of the plant safety framework, common data component is proposed to manage all common data for the plant safety framework and to accumulate the safety-related common data across plant activities. This component is a part of the data warehousing for the plant-wide activities across the plant lifecycle. The object-oriented modeling approach is used to abstract these common elements within the plant-wide conceptual model while the physical data are within the data warehouse frame. The safety common data (SCD) component includes (but not limited to): the possible source of data errors, documentation standards (vocabulary), generic cause-consequence for each component type, checklists for operation-type jobs, and standard safety interlock levels. The common data are essential to be organized and formatted in a standard way to enable smooth utilization by all plant lifecycle components in alignment with the plant-wide data warehouse format. Example of possible sources of data errors is shown in table 3-2.

Table 3-2: Possible source of Data Errors

Data Source	Possible Error	Corrective Actions
Field Measurement	Noise	Filters
	Process upset outside sensor span	Limit check with alarm
	Instrument failure	Limit check with alarm Limit check with default Limit check with last good value Multiple field instruments with signal selection Data reconciliation by gross error detection
	Loss of field power	Hazard zero detect with alarm
	I/O system failure	Integrity check with alarm
Operator input manual data entry	Bad data Key entry error	Interactive limit check with alarm
External data system serial communication link	Bad transmission	Limit check with alarm Message validity check Limit check with default
	Bad data	Limit check with alarm Limit check with default
Application software	Program error	Rigorous testing Limit check with alarm
	Input data limits allow invalid calculation	Limit check with default

3.4.9. Safety Scenarios Component

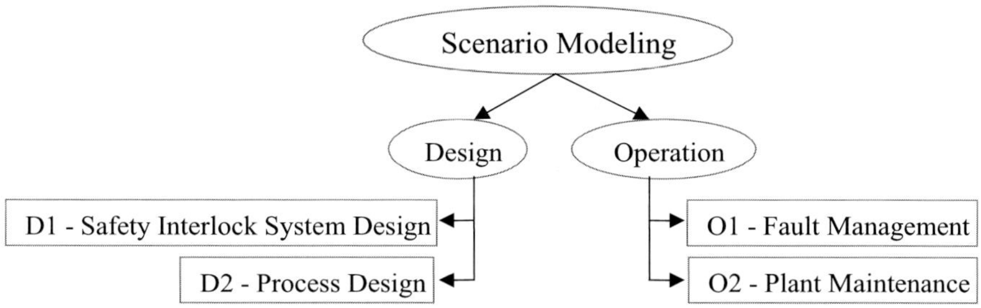

Figure 3-12: Scenario Component Utilization

Scenario is a sequence of events that can lead to a desired event starting from an initial point. Usually the scenario development needs to follow logical and optimum paths using a defined method to ensure the correctness and optimality of the scenario. Safety scenarios are those paths starting from an initial event till it reaches final events going through different events and states of the process. These events are mainly either normal situation or abnormal situation. The proposed safety scenario component could be used for four main objectives: Safety Interlock System Design, Process Design, Fault Management, and Plant Maintenance as shown in figure 3-12.

The scenario component is capable to store pre-defined scenarios and to maintain them (with efficient algorithms in a minimum time) subject to any change in the process. Also, the scenario component is capable to cope with the real time process operations to reflect up-to-date status of the plant and to develop different scenarios for any desired path. ETA (Event Tree Analysis) (CCPS, 1993) is proposed to develop the different scenarios for D1 and D2 algorithms, while FTA (CCPS, 1993) is used to develop the scenarios for O1 and O2. The major difference between the FTA and the ETA is that FTA focuses only on the components faults starting from the top event (major component or process) then goes into sequence of possible faults till it reaches the smaller subcomponents or parts, while ETA considers both the success and the failure starting from initiating event. ETA has mainly two types: pre-incident, which evaluates the effectiveness of the multi-element protective system, and post-incident, which is commonly used to quantify the various consequence types that might arise from a single release of hazardous material (CCPS, 1993). The scenario builder module is proposed to generate all possible scenarios associated with each plant component. Such module can be integrated within the plant safety framework. Using the scenario builder, list of possible (and critical) failures that may cause hazard can be generated. The class model describing the safety scenario component can be viewed in figure 3-13, which reflects the relationship between fault and failure through the link between process variables and process components, while each component can possess normal and abnormal behaviors.

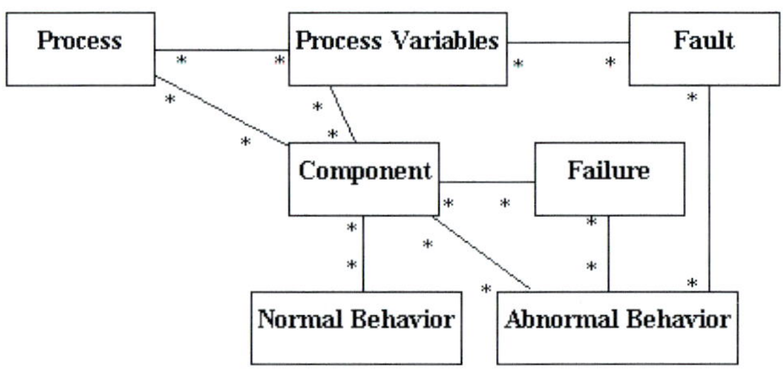

Figure 3-13: Scenario Component Ontology Model

Usually the HAZOP study is used to generate the top events for the analysis tree especially in the initial plant design stage. In order to construct the FTA for the possible fault scenarios the plant structural class model has been constructed, which explains the inheritance from the top structural element (metaclass) "Plant Structural Entity" up to the smallest physical structural components (i.e. valve), indicating the association of the possible failures or faults with each model element, as shown in figure 3-14.

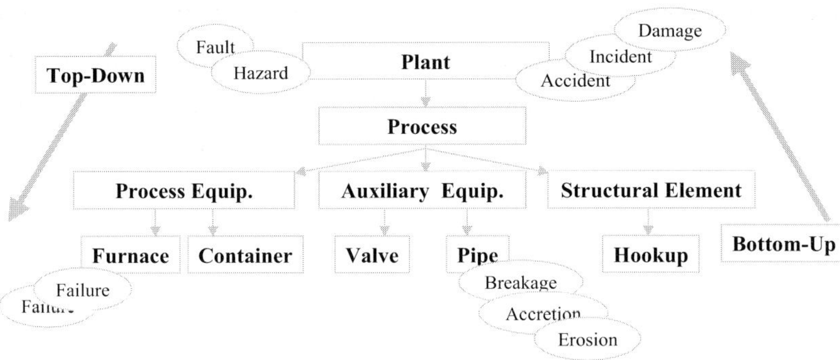

Figure 3-14: Failure Inheritance in Plant Physical Model

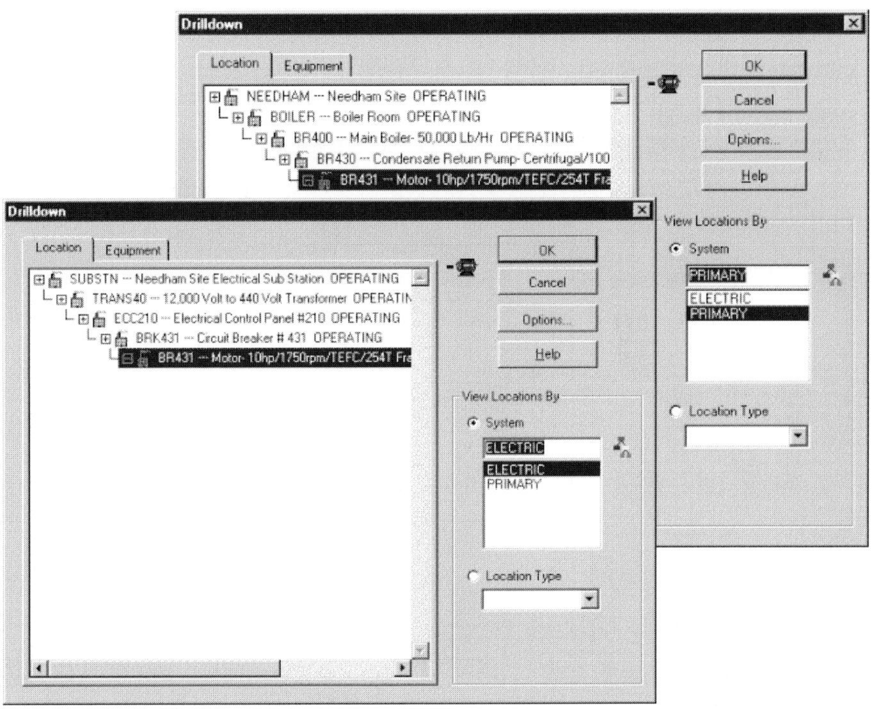

Figure 3-15: Plant Structure Abstraction

The plant equipments (physical components) are organized hierarchically on the basis of the system or equipment type (views) as in figure 3-15 where motor is considered as a part of the electrical system as well as the primary system. The faults also are inherited from all views upward and downward, where FTA and ETA are generated for all cases and stored within the scenario component of the safety framework. As for the hazardous materials used within each component, it is considered to tune the FTA/ETA and to eliminate unrealistic scenarios. This has been expressed by defining the list of hazardous materials composing each component of the plant as well as the hazardous materials that is used with each component or part including the material reference ID, description, MSDS, health rating, flammability rating, and reactivity rating, as shown in figure 3-16. Other fields can be added to the model as needed and as per the type of the plant and material used. The equipment hierarchy based on systems and location types (multiple views) is implemented using computerized maintenance management system (MAXIMO).

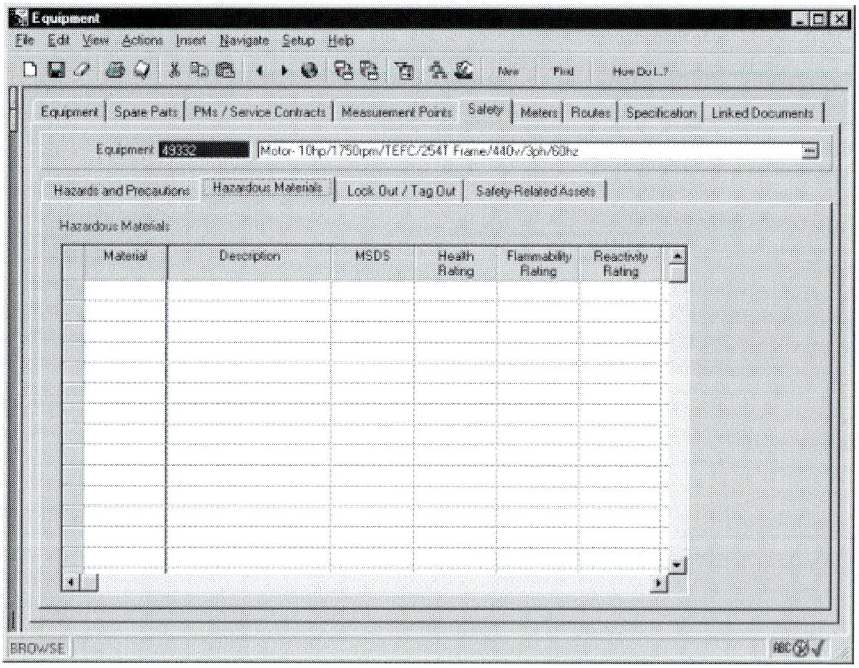

Figure 3-16: Hazardous Materials Associated with Plant Equipment

In order to automate the generation of the FTA/ETA from the plant model during the design and operation stages, an automated FTA/ETA application is proposed using Cara (Cara™) integrated with G2 (G2™), which enables the user to graphically input the plant design using Plant Model Editor and generate the FTA/ETA associated for the selected action. Once the integration between G2 and MAXIMO is implemented it will be possible to correlate the plant equipment list data and synchronize them between maintenance and plant design and operations sides. The generated ETA/FTA will be stored with each plant component and can be modified (or regenerated) upon any change in the plant process. The normal sequence of plant corrective maintenance starts after the occurrence of any failure or fault in a process using the control devices to specify the faulty equipment. However, predicting the faulty component/part within the equipment needs more analysis and historical data to assist in the prediction. For that, ETA/FTA has been proposed to analyze the erroneous equipment and to predict the faulty component. This can be linked to the maintenance job to specify the simplest way to repair the equipment with the help of the safety procedures or jobs associated with the maintenance jobs. By this way, the maintenance cost, time, and effort can be reduced while sustaining the process safety level. Upstream end of an oil refinery process (Ogunnaike, 1994) shown in figure 3-17 is utilized to express the safety scenario component (explained above). The process variables are identified as "T" for temperature, "F" for flow rate, and "P" for pressure. The F* and T* are identified as the desired output process variables to be monitored and controlled (the stability and controllability of the product output rate is critical to fulfill the market demand). For that plant objects are defined within G2 as inherited objects from the top object "ITEM". The Fault Tree generated by G2 is exported into CARA for conducting the analysis and quantifying the results based on the defined parameters.

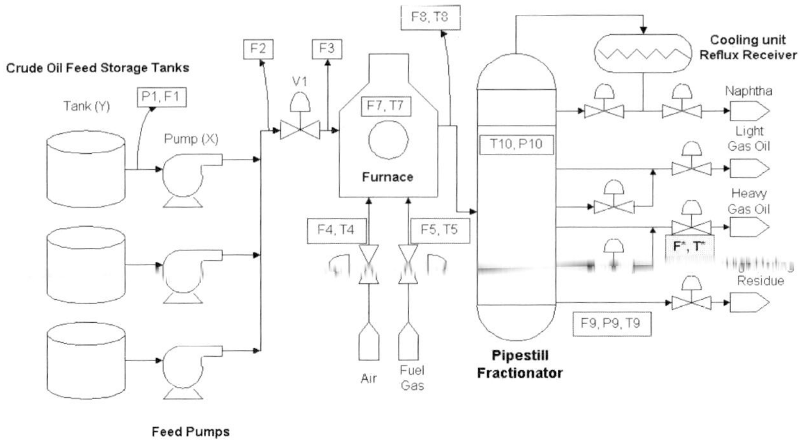

Figure 3-17: Upstream end Process of Oil Refinery Plant

The fault tree of the upstream end of the oil refinery that is developed within CARA is shown in figure 3-18. The top event for the fault tree is selected to be F* is low, which implies that the flow rate of the output product is low (any of the specified products – Naphtha, Light Gas Oil, or Heavy Gas Oil). The result of the fault tree analysis as generated by CARA shows total of six cut sets up to order four. The results of the fault tree are fed back to G2 environment to be linked with the plant model and associated with each plant physical object within the safety scenario component. The relationships among process variable, fault, component, and failure help in establishing the basis for the scenario component. The minimum cut sets generated by CARA represents the different scenarios for the occurrence of the top event (output product flow rate is low). These prepared safety (or fault) scenarios are stored along with each associated component to enable the plant environment to manage the faults efficiently.

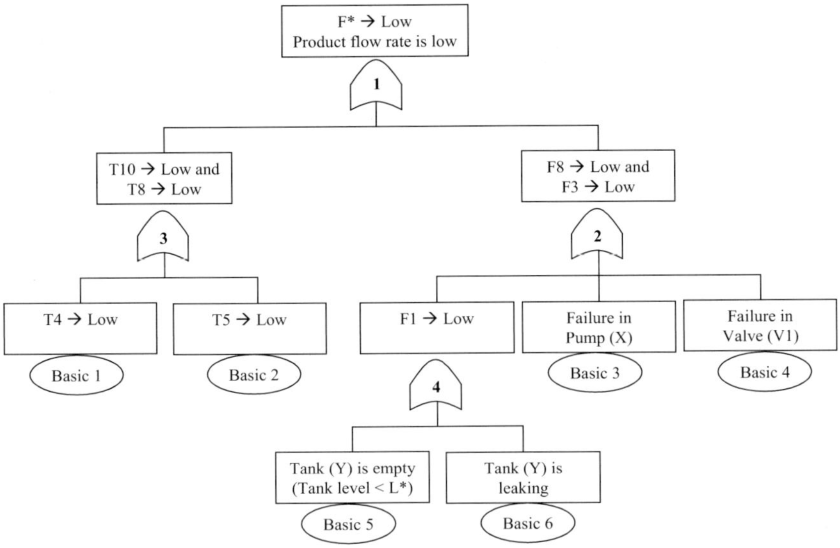

Figure 3-18: Fault Tree Analysis for Oil Refinery Using CARA

3.4.10. Safety Devices Component

Process plants need to utilize automatic control systems not only to enhance process safety, but also to improve equipment availability, achieve high quality product, and to minimize human interference. There are two major types of process safety devices, first is to support the normal production functions (referred as basic process control systems BPCS), while the other is dedicated for protective functions like monitoring key process variables, solving process state logic, and managing abnormal situation by assisting the operator (CCPS, 1992). Such protective controls are referred as plant safety interlock systems (SIS). Most of the textbooks and researches associate process control with process safety where process control assures process safety. Plant development starts with the requirements definitions followed by evaluation of production technologies taking into considerations controlling criteria (economics, reliability, regulatory and corporate requirements, acceptable risk, etc.). Selecting the preferred process technology includes identifying hazards and conducting quantitative/qualitative risk assessment. Plants with potential to cause hazard should be constructed with multiple independent protection layers, which is designed to avoid and/or mitigate the harmful effects of hazards (CCPS, 1992). The proposed safety control systems component manages the two types of safety devices (BPCS & SIS). The proposed architecture of the safety control systems components can be viewed in figure 3-19 (CCPS, 1992). The safety interlock systems are designed to formulate the independent protection layers (IPL) as shown in figure 3-20 (CCPS, 1992). The design and locations of the different components of these protection layers can be determined using dynamic simulator with process design and variables as input, while the specifications of the control devices are the expected output, which can be developed using G2 connected to process design in CAD environment. A prototype dynamic simulator with safety evaluation system has been developed within our lab using G2 to identify the locations of the protection layers based on the calculated risk using the estimated probability by introducing the unavailability of each event via Dow's F&EI (Chikaraishi and Suzuki, 2000).

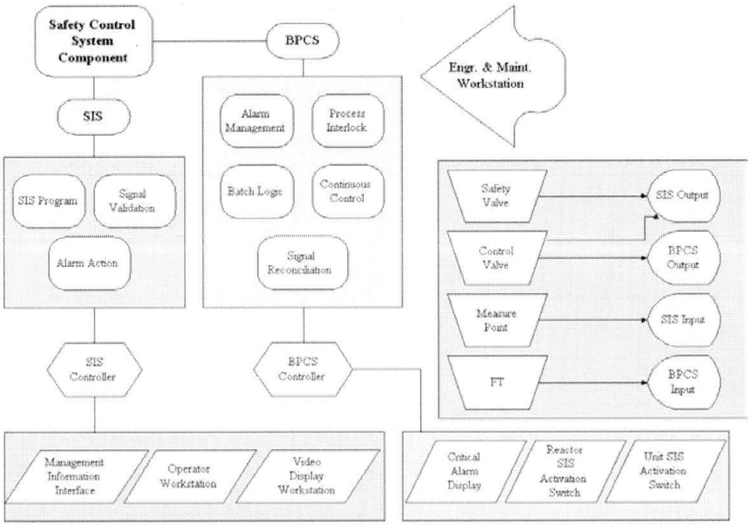

Figure 3-19: Safety Devices Component – Safety Control Systems Architecture

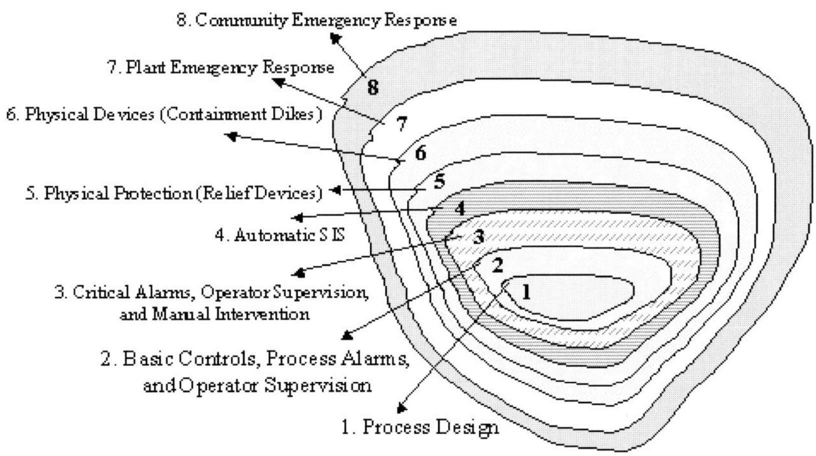

Figure 3-20: Typical Protection Layers (IPL) in Modern Chemical Plants

3.5. Fault Propagation Modeling

There are different methods that are currently used to model the fault propagation within plant processes such as sign directed graphs, or simply SDG, (Androw and Lees, 1975). One major limitation of sign directed graph is its limitations to represent the environment factors that affect physical equipments.

One key issue to automate the hazard evaluation process is to find a systematic way to represent the fault propagation in the plant process. This section describes a new way of representing the fault propagation in process plant using examples from HDS reactor CGU.

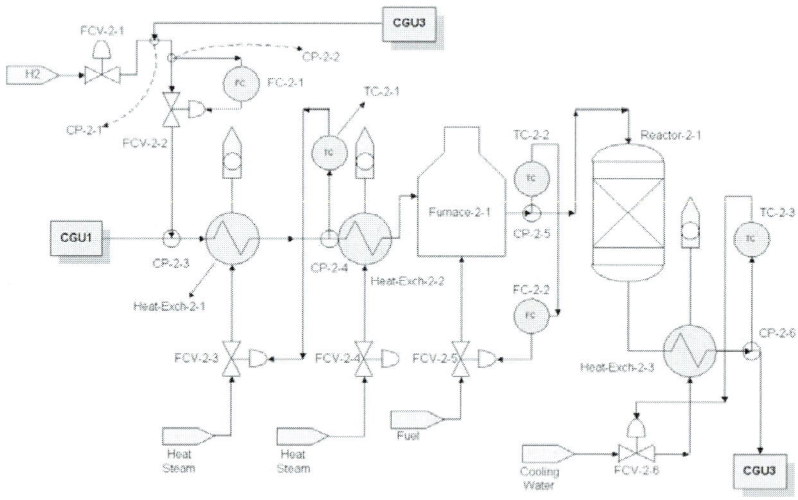

Figure 3-21: Simplified P&ID of Reactor Process of HDS Plant

Reactor CGU of HDS is used as a case study to illustrate the fault propagation model. The simplified P&ID of the reactor CGU is shown in figure 3-21. The equivalent object-oriented model representation of the reactor CGU is shown in figure 3-22. In this diagram, each component within the P&ID is represented as a class, while the fluid flow among the different components is represented as dataflow with stereotype of the fluid class.

The approach proposed here is developed by abstracting a list of fault propagation statements generated by experts during the hazard evaluation process while performing cause-consequence analysis of the selected case study (reactor CGU of HDS plant). Table 3-3 shows sample of the cause-consequence analysis of reactor CGU from HDS plant. In this table, the equipment are numbered as per the numbering scheme followed while representing the plant model using the proposed object-oriented modeling approach.

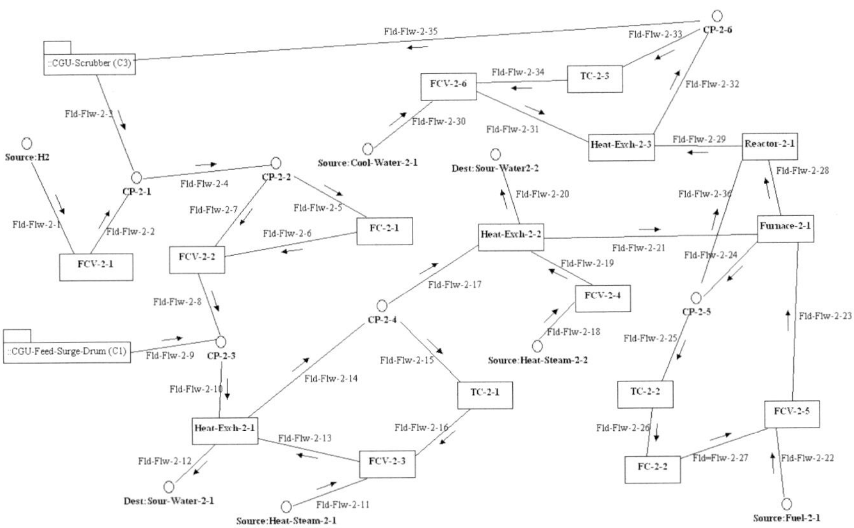

Figure 3-22: Reactor CGU Model Representation in UML

The scenario shown in this table presents the initial event of failure in the temperature control valve TC-2-2, where the temperature is lower than expected.

3.5.1. Scenarios from Cause-Consequence Analysis

Table 3-3: Scenario (S1) – Cause/Consequence Analysis for Reactor Unit of HDS Plant

Cause	Consequence
(1) Failure of TC-2-2, Temperature low	(2) Feed amount of fuel Gas to Furnace-2-1 becomes bigger
(2) Feed amount of fuel Gas to Furnace-2-1 becomes bigger	(3) Abnormal combustion inside Furnace-2-1
	(4) Temperature increase of the liquid mixture Fld-Flw-2-28 (Diesel + H2)
(3) Abnormal combustion inside Furnace-2-1	(5) Breakage of Furnace-2-1
(5) Breakage of Furnace-2-1	(6) Thermal emission to outside (disturbance)
(4) Temperature increase of the liquid mixture Fld-Flw-2-28 (Diesel + H2)	(7) Temperature increase in Reactor-2-1
(7) Temperature increase in Reactor-2-1	(8) Temperature runaway inside Reactor-2-1
	(9) Temperature increase of catalyst (H2) in Reactor-2-1
(8) Temperature runaway inside Reactor-2-1	(10) Fracture of Reactor-2-1
(9) Temperature increase of catalyst (H2) in Reactor-2-1	(11) Possibility of coking in Reactor-2-1
(11) Possibility of coking in Reactor-2-1	(12) Blockage around nozzle
	(13) Loss of pressure in the Reactor-2-1 by coking
(12) Blockage around nozzle	(14) Flow rate decrease of H2 for quench
(14) Flow rate decrease of H2 for quench	(15) Temperature increase in Reactor-2-1 (loop back to 7)
(13) Loss of pressure in the Reactor-2-1 due to coking	(16) Flow rate decrease of H2 due to quench
	(17) Temperature increase of catalyst H2 in Reactor-2-1 (loop back to 9)

3.5.2. Fault Propagation Layers (FPLs)

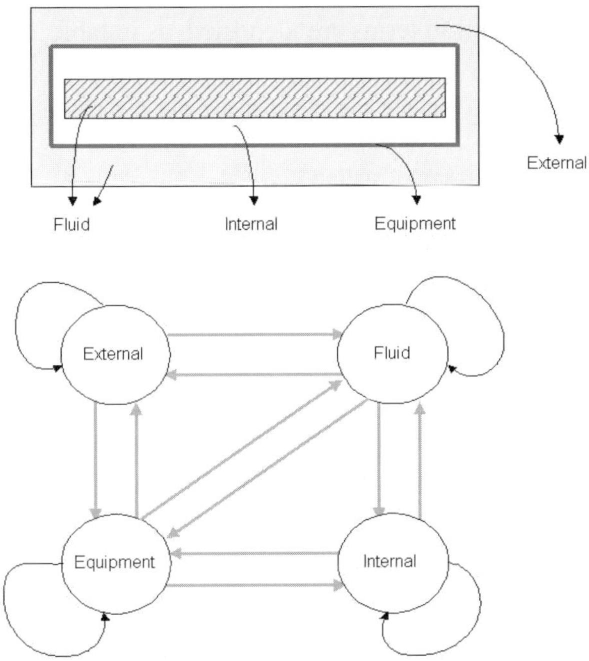

Figure 3-23: Fault Propagation Layers State Transition Diagram

In order to understand the mechanism of failure propagation it is essential to define abstract layers in which the fault can propagate, which is called Fault Propagation Layers (FPLs), as shown in figure 3-23. In this figure, there are mainly four failure propagation layers within the plant process: Fluid, Internal, Equipment, and External, as follow: the first layer could be the fluid flowing inside the equipment; the second layer is the internal environment within an equipment (which includes mixture of fluids); the third layer is the equipment itself; and the fourth one is the environment external to the equipment (which includes other fluids like air, other equipments, and/or other objects like human, buildings, sea, etc). Any of these layers can contribute in the failure propagation (inwards or outwards).

3.5.3. Fault Propagation Modeling

By applying the FPL model to the above cause-consequence results, the following abstraction terms are identified, as in table 3-4.

Table 3-4: Process Failure Abstraction

Scenario Step	Physical Equipment Class (PEC)	Normal/ Abnormal Situation (NAS)	Process Variable (PV)	Fluid (FLD)	Operation (OPR)	Position in Process Equipment (POS)
S1-2	Furnace-2-1	Amount Big	Amount	Fuel Gas	Feed	Inlet
S1-3	Furnace-2-1	Abnormal Combustion		Fuel Gas		Inside
S1-4	Furnace-2-1	Temp Increase	Temp	Liquid mixture (Diesel + H2) Fld-Flw-2-28	Mix	Inside
S1-5	Furnace-2-1	Breakage			Mix	Body
S1-6	Furnace-2-1	Thermal emission		Liquid mixture and gas	Mix	Outside

By analyzing the above table, the combinations of the different columns can be summarized as per table 3-5.

Table 3-5: Fault Propagation Result Analysis

Physical Equipment Class (PEC)	Normal/ Abnormal Situation (NAS)	Process Variable (PV)	Fluid (FLD)	Operation (OPR)	Position in Process Equipment (POS)
√	√	√	√	√	√
√	√		√		√
√	√	√	√	√	√
√	√				√
√	√		√	√	√

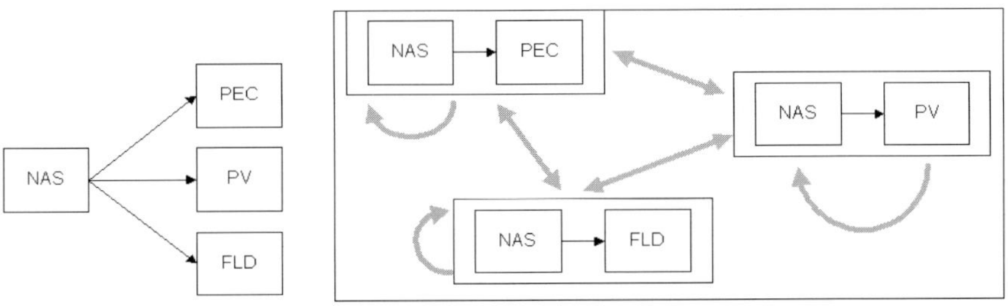

Figure 3-24: Fault Propagation Schemes

The idea behind this analysis is to detect a certain pattern that can be used to automate the fault propagation modeling and analysis practices. It is obvious that for each fault description, there should be "NAS", "PEC", and "POS". However, the NAS could be related to equipment i.e. Breakage, or related to fluid i.e. "Big Amount" of fluid, or related to operation i.e. fail to close valve. Figure 3-24 shows the possible fault propagation schemes, which shows that NAS can affect PEC, PV, or FLD. And NAS affecting PV, FLD, and PEC can propagate to each other. For example, failure in equipment can lead to abnormal situation in the fluid, internal, or external environment, while abnormal situation in fluid within the reactor can lead to failure in the internal environment of the reactor or the reactor itself.

3.5.4. Fault Propagation Ontology

From the above analysis, we can define the ontology model of the fault propagation based on: NAS, PEC, PV, FLD, OPR, and POS, as in figure 3-25.

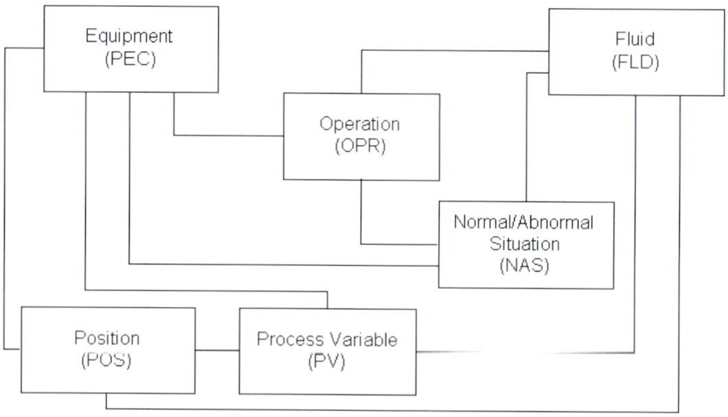

Figure 3-25: Fault Propagation Ontology Model

3.5.5. Fault Propagation Model Representation in POOM

The above analysis and model is useful to enable us to draw the structure of the database of the failure data group and to automate the hazard evaluation practice. But in order to do that it is essential to define the fault propagation model and its representation in the plant design model using the above analysis results.

Fault propagation model elements (FPMEs) are as stated earlier: PEC, PV, NAS, POS, FLD, and OPR. These elements can be mapped to the plant object oriented model (POOM) as per table 3-6.

Table 3-6: Fault Propagation Model Mapping to POOM

Fault Propagation Model Element (FPME)	Representation in POOM
PEC	Physical equipments are represented as object, class, or package. If PEC represents a complete CGU then it is mapped to package, otherwise it is mapped to object or class.
OPR	Operation is mapped to the operation, which is associated with class or object. Operation can accept parameters to perform the step-by-step functions.
PV	Is a class variable, which is associate with the equipment class.
NAS	Is normal or abnormal situation, which is an events or a state, which represents the failure mode i.e. leaking, or abnormal situation like temperature is high.
POS	The position of the NAS is related to the running equipment, affected fluid, and the process variable. POS can be represented as class variable associated with equipment, fluid, or as a sub-class associated with the equipment or fluid.
FLD	Fluid can be represented as class, which can be used as a stereotype to be associated with the fluid flow from point-to-point in the plant design model.

Faulty states can be identified in the state diagram as part of the behavior model of the plant. The transition from normal state to faulty state can represent the failure mode. The transition from faulty state to another faulty state can represent the fault propagation model showing the different scenarios. The different scenarios of fault propagation paths for each process can be represented as state diagrams associated with the process. The safety literature highlighted the use of *Sigh Directed Graphs* (SDG) to represent and analyze the fault propagation within a plant process (Umeda, 1980). For a given process, SDG can be modeled within UML plant model. SDG can be defined for each plant physical object. SDGs representing plant components can be connected to form the overall fault propagation model that covers plant physical objects. The in-flow and out-flow are considered as logical objects within the plant static dimension. The increase or decrease of the flow (or process variable) is a conditional branching (transition) within the state diagram. The transition from in-flow to out-flow can be represented as messages (signals) with "+" or "-" as an associated information. Figure 3-26 shows a sample fault propagation model of a control valve. The proposed corrective action can be viewed as transition from the faulty state back to the normal state. Such information is obtained from the plant lifecycle knowledgebase that intelligently diagnose the fault and proposes the most suitable corrective action.

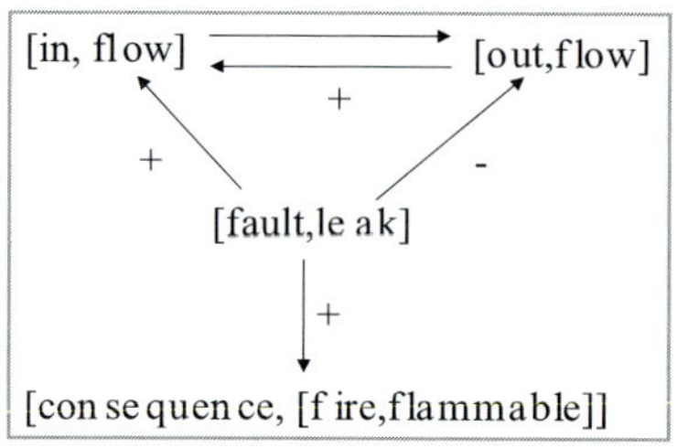

Figure 3-26: Fault Propagation Model for Control Valve Represented as SDG

3.5.6. Fault Propagation Automation using Knowledge Engineering

Fault propagation knowledge can be represented in the form of two groups:

- Fault Propagation Model Element (PEC, PV, etc.)
- Fault Propagation Rules (which is used for reasoning in the inference engine)

There are several approaches proposed to analyze the fault propagation i.e. SDG. Knowledge engineering is one proven approach to automate the analysis fault propagation of complex plants where knowledge is accumulated over many fault exercises and on many plants. This approach can be easily automated using a computer program, which will store list of all possible values associated with FPMEs while maintaining a list of rules to relate these FPMEs. Neural Networks is one suitable approach to enhance the learning of such expert system.

Part II: Plant Enterprise Engineering Environment (PEEE)

4. Plant Enterprise Engineering Environment (PEEE)

The term of plant enterprise engineering environment (PEEE) came from idea of integrating both the engineering and non-engineering environments to share plant lifecycle data. Currently, most of the plants have two major divisions within the organization, where the engineering systems, i.e. design and operation, are isolated from the non-engineering systems i.e. procurement and financial systems. It is proven that the integration between these two domains is essential to optimize the engineering activities. One major example is the utilization of maintenance history and reliability data (from the non-engineering side) in design rationales (within the engineering side). In order to achieve this picture it is essential to address technology related issues as well as information ones. The ultimate goal of having such integrated environment is to manage both the engineering and non-engineering activities and to enable the communication with the outside world i.e. supplier, customer, and other plants. This section describes complete picture of such environment and resolve some issues related to both sides and discusses ideas about how to achieve such environment.

4.1. PEEE Functional Analysis

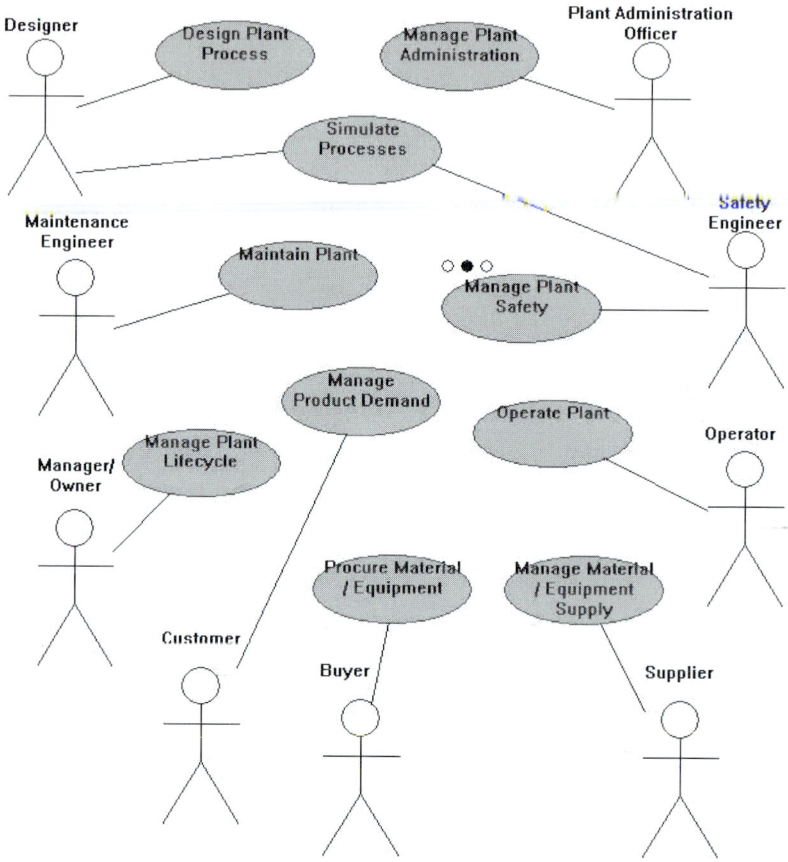

Figure 4-1: PEEE Use Case Modeling

Prior to defining the technology and information system architectures, a functional analysis has been carried out using Use Case modeling to identify the major functions and roles within the plant enterprise. In the top level Use Case modeling of PEEE, the key players or roles as well as the main functioned offered by PEEE are illustrated in figure 4-1. In this diagram Use Case modeling has been conducted using OO CASE (System Architect 2001™). Among the main roles in the plant enterprise we can find designer, operator, owner, safety engineer, safety engineer, and maintenance engineer. The main functions offered by the integrated environment are: maintain plant, operate plant process, manage plant safety, and design plant process. As the three dots associated with the use case "manage plant safety", this means there are more detailed use cases, or diagrams associated with this use case. This top down approach enabled us to reach the detailed level of object-oriented analysis of PEEE, towards the understanding and reengineering of the PEEE process and to develop the integrated systems required to cover PEEE.

The high level modeling of PEEE helped in identifying the major components and integration links. In addition to the Use Case modeling, the main domains of the system have been identified and represented within POOM as packages, which includes: procurement, operation, financial, human recourses, safety, design, and others. From these models, CAPE-SAFE roles, use cases, and integration with other components and humans within PEEE is identified.

4.2. Information Technology Infrastructure

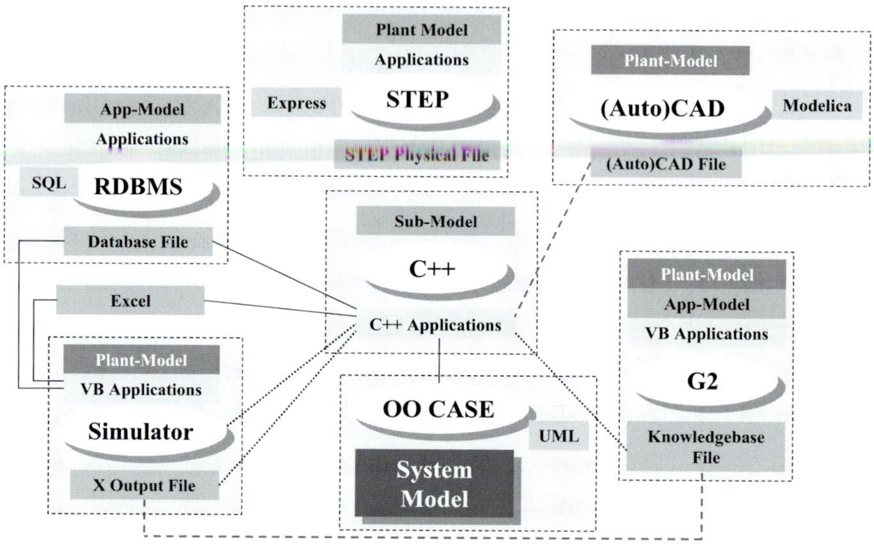

Figure 4-2: Example of Information Technology Infrastructure of PEEE

Figure 4-2 shows an example of the current PEEE information technology infrastructure where models and systems are distributed over different domains using different standards, where in each domain contains part of the plant model, sub-system, tool, output file, and standard followed. By analyzing the current infrastructure (as in figure 4-2), and restructuring the components, a new PEEE information technology infrastructure is proposed as in figure 4-3.

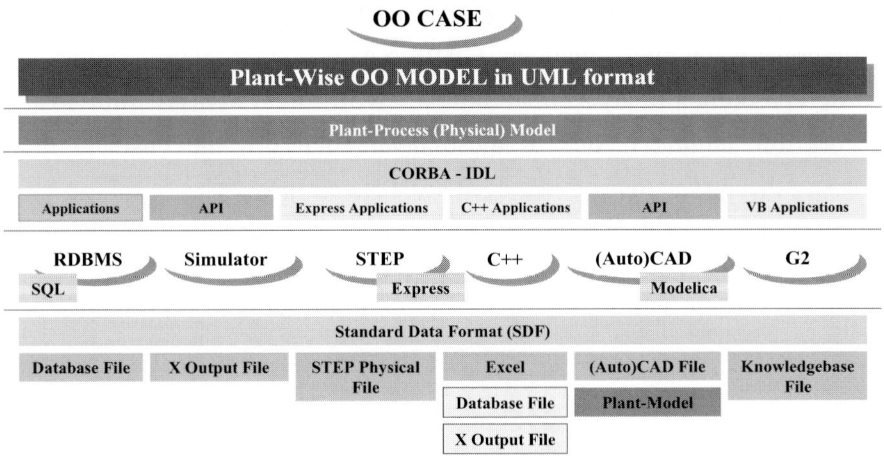

Figure 4-3: Proposed PEEE Information Technology Infrastructure

In the proposed technology, the middleware layer has been proposed i.e. CORBA to integrate the different systems and APIs. i.e. component-based software systems (following CAPE-OPEN standards). Utilizing the middleware in PEEE will improve the system development and maintainability, which will reduce the cost and efforts required to develop PEEE, in addition to assuring the integrity and consistency among the different systems and applications. Such environment makes it easy to add a new feature or service for PEEE, using the developed component library. The different output files are recommended to be in a standard data format (SDF), which enables any tool to present the output data via standard user interfaces. This picture has impact on the designed CAPE-SAFE solution, especially in the integration with design and operational systems.

4.3. PEEE System Architecture

Figure 4-4 shows the system architecture of PEEE, which includes the different systems that manage the plant lifecycle activities. Health, safety, and environment management system (HSEMS) is one of these components, which includes CAPE-SAFE. This figure shows the different common layers that interact with all systems within PEEE. Workflow engine is one of these layers that ties the business layer among these systems and manages the business processes.

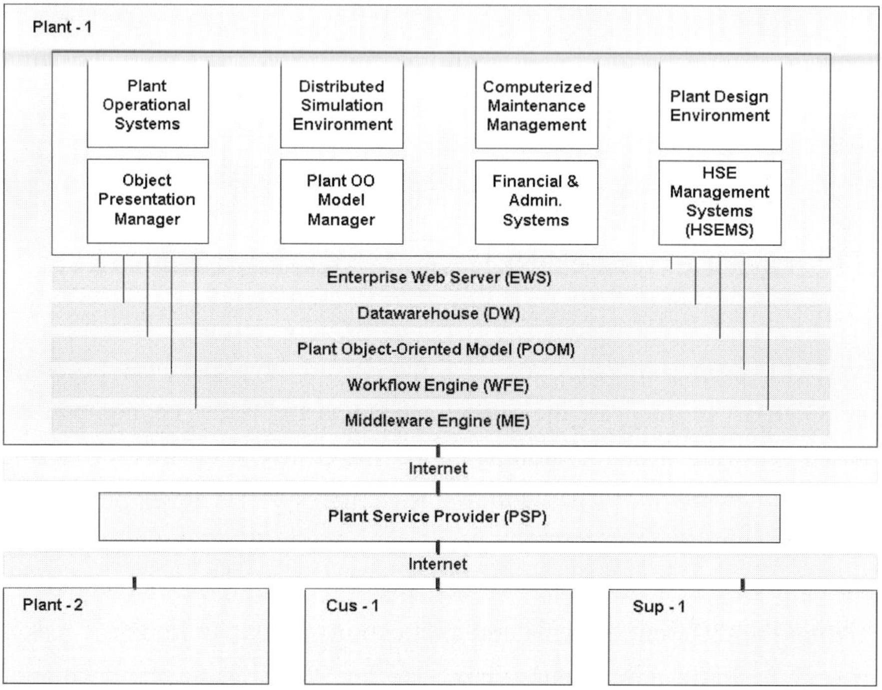

Figure 4-4: Plant Enterprise Engineering Environment System Architecture

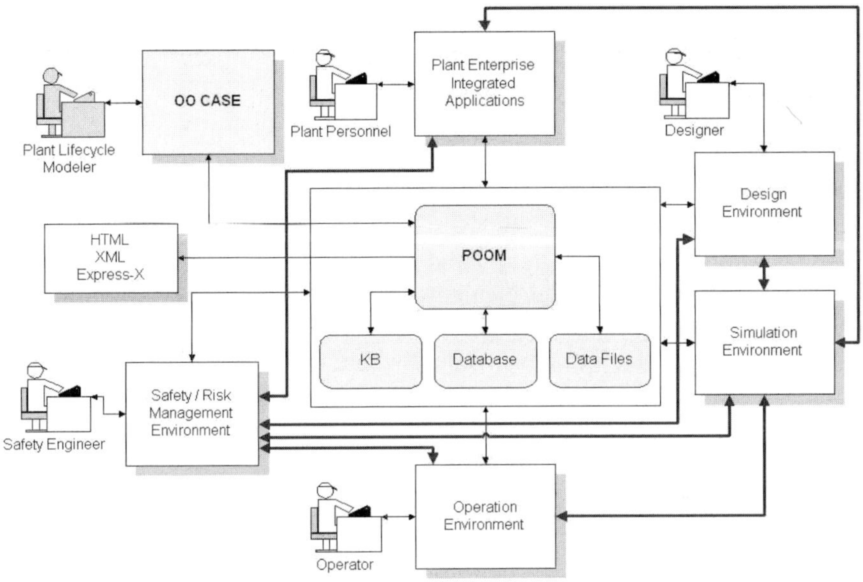

Figure 4-5: Plant Enterprise Engineering Environment Flow Chart

Figure 4-5 shows the system flowchart, which comprises safety, design, simulation, operation, ERP, and modeling environment, with safety engineer, operator, designer, modeler, and plant personnel as roles within the global picture of PEEE. In this figure, Safety engineer will be interacting with the safety management environment (CAPE-SAFE), while operation, design, simulation, and plant enterprise applications (ERP) will be integrated with CAPE-SAFE.

4.4. PEEE Components

As part of design phase of PEEE, it is essential to identify the major components as shown in table 4-1.

Table 4-1: PEEE Components Description

CAPE-ModE	It is the automated modeling environment that manage the model elements of the different models within the plant enterprise.
CAPE-CAD	Design environment where designer will be working to design plant process.
CAPE-Sim	Simulation environment which is intended to simulate any process within the plant, i.e. engineering and non-engineering.
CAPE-Oper	Operation systems which include operation support system
CAPE-SAFE	Plant enterprise safety management system
CAPE-ERP	The integrated enterprise applications that manages the resources like maintenance, financial, human resources, etc.
CAPE-Train	The training module that provides and manages training in different disciplines of the plant, i.e. safety, operational, and maintenance training.
CAPE-PSP	Plant Service Provider, is a new proposed idea to remove the barriers between plant, supplier, and customer by providing means of communication and sharing data.
CAPE-Manager	The manager program that is used to manage the different modules and systems.
CAPE-WF	Workflow manager and organize the business processes.
CAPE-Admin	It is different than the manager program, as it enable data administration, and security of information. Also it provides facilities to configure databases and look and feel.

4.5. CAPE-PSP

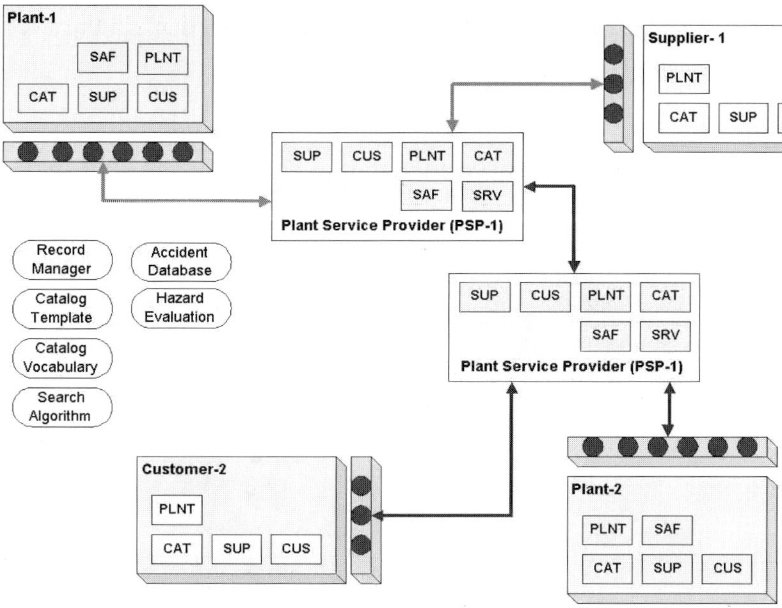

Figure 4-6: Plant Service Provider Design Architecture

Plant Service Provider is a new concept introduced to enable plants to communicate with each other, with suppliers, and with customers. This includes providing and registering services that can also be shared among the different parties. The implementation of this facility will enable plant lifecycle data sharing and will facilitate the supply chain and demand chain management. Figure 4-6 shows a proposal of the mechanism of PSP where plants, suppliers, and customers are connected to the PSP. In this proposal different PSPs can communicate with each other to update their indexes. The Internet media is used as the communication link among the different parties. One example service is the CAT service, which updates the plant catalog system transparently with new catalog items or specifications that may exists in other plants via PSP, who provides information about the location and specification summary of similar catalogs and their contents. The realization of this proposal requires further study in the area of organization, control, and security of such hybrid systems. Agent technologies and approaches can offer suitable approaches to implement such solution.

Plant Safety can be enhanced in different aspects using PSP, where accident database can be updated transparently among the different plants, also hazard evaluation results for similar processes can be shared, searched, and exchanged. One major advantage is the communication between the engineering company and the operating company through PSPs to update or synchronize plant design and safety related information.

5. Plant Modeling Environment (CAPE-ModE)

The need to describe the plant enterprise modeling environment within the context of this book is to define how to manipulate the model elements related to thc safety model within the plant lifecycle. This will specify in more details how CAPE-SAFE manipulates the plant model elements related to plant safety aspects.

Throughout the plant lifecycle there are many models that maintain certain view(s) about the plant lifecycle. The model of one stage could be part of the user objects in other stages within the plant lifecycle. For example, plant design model elements representing the P&ID can be used to formulate operation, safety, and maintenance models. This means that the model elements of the plant design model are in fact metamodel elements for the plant operation model.

The intended automated modeling environment will provide standard and central pool where model elements are represented in unified modeling language and manipulated using the central model management environment with the use of intelligent user interface.

The practicality and applicability of constructing such integrated model in a systematic and standard manner is the prime concern to achieve the plant enterprise engineering environment. The target model need to satisfy general features and characteristics like reuse of its model elements and flexibility for expansion as well as standard representation in user-preferred formats like HTML, XML, or Express-X. The model formalization, utilization throughout the plant lifecycle, and the design of the computer-aided modeling environment (Marquardt et al., 2000) are three major subtasks towards achieving the main objective. The focus in this section will be on designing the computer-aided modeling environment that can support the model formalization and utilization by other systems and components within the plant enterprise i.e. CAPE-SAFE. In general, lifecycle-centered view on the development process focuses on tracking the evolution of an artifact (such as a chemical plant) from its initial creation, through a number of reengineering cycles till it is finally discarded (Bayer et al., 2000). STEP has taken positive initiatives to integrate and represent the product lifecycle data aiming to standardize the data formats so that it can be easily exchanged (electronically) among the different tools. Such approach requires building the plant lifecycle physical data models that represents the structure of lifecycle data. STEP Tools Inc. has devoted its efforts to build STEP solutions, which includes ST-Developer to build and maintain STEP applications, ST-Repository to manage STEP product databases, ST-Viewer to view 3D CAD information, and group of translators to translate to and from STEP data files (found in the reference URL of STEP Tools). STEP strategic vision towards enterprise information management dictates the necessity to continue the work to develop more detailed STEP libraries and automated tools to support such data-centered approach to manage enterprise information. A complementary vision (and could be an alternative) to such approach is the model-centered view where the focus will be to construct higher-level in abstraction as a building blocks. These building blocks or classes will be used to construct the plant integrated model. Such hierarchical construction of the plant model has a major difficulty of having large number of classes. This problem has been solved using the concept of metamodeling. One definition of Metamodeling is described by Linninger (1999), as it is a system architecture that supports custom-built modeling languages. Its main conjecture is user adoptability. It encourages the free institution of specific modeling languages. This concept has been adopted by many research projects (trends have been explained by Marquardt, 1996) for instance RWTH Aachen has developed an automated modeling environment ModKit (as presented by Bogusch, 1997) that supports the process design and simulation modeling. Such modeling environment has been integrated with process simulator system called “Cheops”, model repository ROME (Wedel et al., 2000), and context-oriented process support “Cops”, which captures existing experience in an activity model and uses this kind of information to guide the modeler to carry out his design work. Such environment allows storing the different models in a neutral representation while physically store the model objects in OODBMS. Full details about the work done by RWTH Aachen (Bogusch et al., 1997, Bayer et al., 2000). So far, OODBMS hasn’t proven any remarkable success to manage large number of objects, this can explain the reason why almost all manufacturers and plant enterprises are still adopting RDBMS with facility to support objects to manage their enterprise information (i.e. ORACLE 8i).

Another example for the modeling environment called “SELF” has been explained by Agensen (1993). SELF features a class-less inheritance mechanism and proposes the separation of data and methods. While the object data are inherited in the traditional object-oriented. In such environment method-object concept has been introduced.

The definition of the automated modeling environment in the problem domain is to assist in constructing and managing all abstract definitions i.e. metamodel elements as well as model elements throughout the plant lifecycle and to provide the means to utilize and manipulate the developed layers of abstraction. The proposed solution is called computer-aided plant enterprise modeling environment or CAPE-ModE. In order to properly design the most efficient plant enterprise modeling environment, it is essential to propose a design architecture of the ideal plant modeler engine as connected with other draw the visionary system architecture of the expected plant modeler engine as connected with other components within the plant enterprise engineering environment.

Figure 5-1 shows that OO CASE can be used as a modeling framework to construct the model repository on the basis of UML. An intelligent user interface is required for other humans and systems within PEEE to manipulate and access the model elements within the model repository. Object presentation manager is a new component proposed to present the object in different ways based on the type of the client, for example equipment class can be presented within the P&ID window as equipment-shape object i.e. icon, while it can be presented as a simple box within the modeling window. The connection to other plants is shown in this figure as it is one of the expected advantages of using standard modeling language i.e. UML, to enable the exchange of the model elements among plants to enhance the process modeling and design. This also can help in plant safety practices where safety aspects are implemented as model elements within the plant model, which can be exchanged among the different plants.

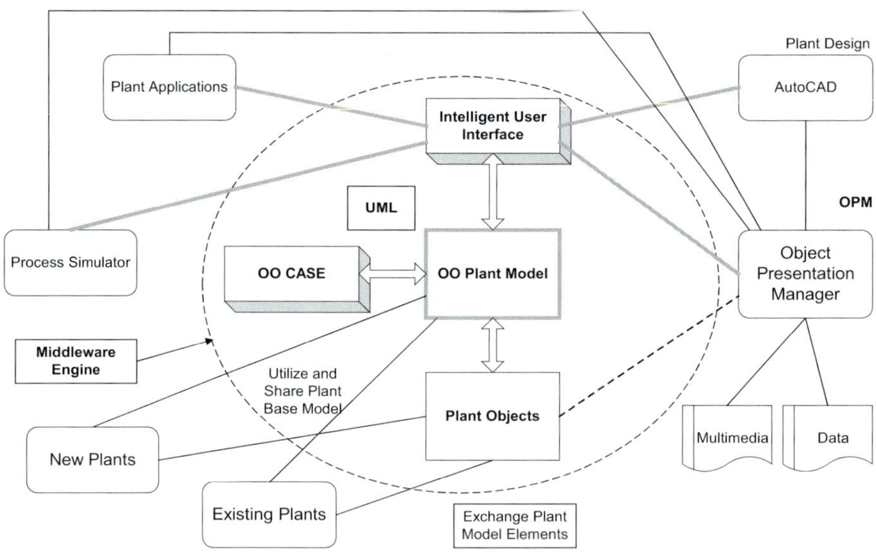

Figure 5-1: Plant Modeler System Architecture

5.1. CAPE-ModE Functional Analysis

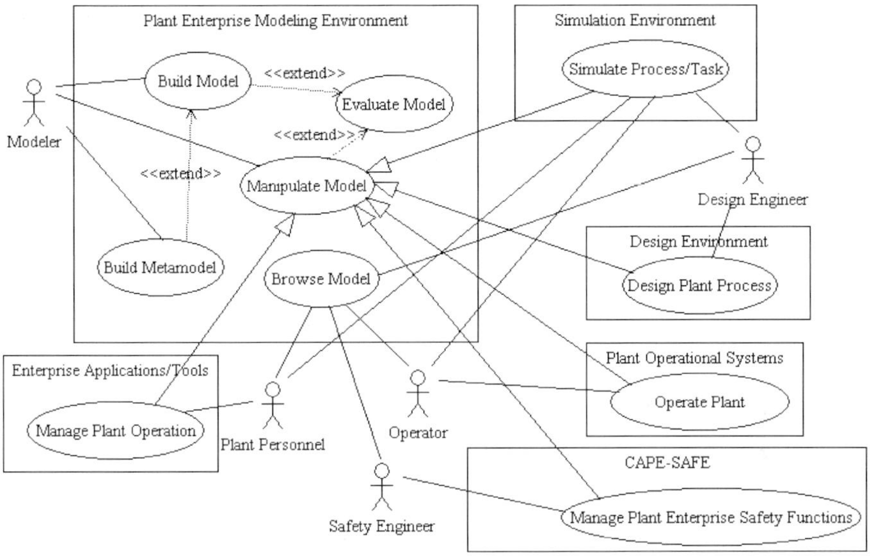

Figure 5-2: CAPE-ModE Functional Analysis using Use Case Modeling

The first step to design the target plant enterprise modeling environment is to identify the actual requirements, which will dictate the detailed specifications to develop such environment. Figure 5-2 shows the use case model as represented using UML notation (UML Notation Guide) that reflects the functional requirements. In this figure, there are four actors interacting with the target modeling environment: (1) Modeler, or the plant enterprise modeler, who is in-charge of maintaining the different models throughout the plant lifecycle, (2) Process designer who will focus on the plant design model elements, (3) Operator who will perform all tasks to operate the plant as per the predefined operating procedures and to keep the plant in safe condition, and (4) Plant personnel who will manage and carry out all administrative tasks and services to operate the plant. Four domains are illustrated as connected to the modeling domain: (i) Simulation domain where all types and tasks of simulation will be performed, (ii) Design environment where process design will be developed and maintained, (iii) Plant operational systems which constitutes the technical systems used by operator to operate the plant, and (vi) Enterprise applications/tools which include administration and service applications and tools to support the operational systems.

The functions (i.e. use cases) covered by the automated modeling environment are as follows;

- Build metamodel: to create and maintain the different metamodel elements interactively.
- Build model: to create and maintain the different model elements defined using the metamodel elements interactively.
- Manipulate model: to maintain the model elements from any external application/tool.
- Evaluate model: as an extended function to evaluate any changes or amendments to the model. This will ensure the consistency of the model throughout the plant lifecycle.

As a complete analysis phase of such project, the above use cases are expanded in more details to reach the detailed functional requirements of the proposed automated modeling environment. From such functional specifications, and from the previous research work, mentioned in the introduction section, there are some major recommendations that to be considered while designing such environment as in table 5-1. The basic components of such modeling environment can be depicted using the above similarities.

Table 5-1: Broad Line Recommendations to Design CAPE-ModE

Similarity to database management system (DBMS)	The concept of database management systems can give us the clue of developing model management system that can manage the model elements within the model repository, which is similar to the database repository.
Similarity to structured query language (SQL)	In order to retrieve the model elements in a standard manner, model query language (MQL) is required. Such model query language can be used via user interface similar to SQL.
Similarity to data manipulation language (DML)	Similar to the DML, model need to be manipulated using a standard model manipulation language (MML), which will be used either within any programming language like C++, or directly within the modeling environment.
Similarity to system tables	Metamodel is the structure that defines the modeling language and used to define the model elements. This is similar to the system tables that define the database structure.
Similarity to import/export facility	To be able to exchange the model elements among different models in heterogeneous environment, import/export facility is required. Such facility will

5.2. CAPE-ModE System Architecture

Figure 5-3 shows the system architecture of the computer-aided plant enterprise modeling environment (CAPE-ModE). The model will be stored within a model repository where both model and metamodel elements will be stored in the form of unified modeling language (UML).

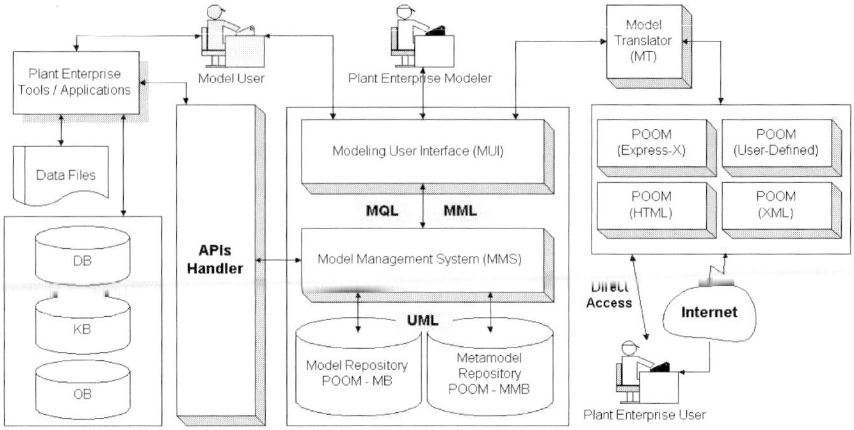

Figure 5-3: CAPE-ModE System Architecture

Model Management System (MMS) is dedicated to manage the model elements in the repository. Such management functionality, which is similar to that of the DBMS, includes security, access, searching, sorting, indexing, etc. User interface will enable the plant enterprise modeler to manipulate the model using Model Manipulation Language (MML) and Model Query Language (MQL). In principle the MML could be similar in functionality to DML, while MQL could be similar to SQL. Intelligent translator will translate the model into any user-preferred format. In this context, XML, Express-X, and HTM are selected as examples of the internationally recognized formats. In order for the different applications and tools within the plant enterprise engineering environment to be able to access and manipulate the model a standard application interface routines are required. Such standard routines (embedded MML or MQL) will be used within the respective source code from the different applications and tools to access, browse, retrieve, and manipulate the model elements within the model repository or model-base (MB). Enterprise users can browse the plant model (POOM) via Internet (in case of HTML output format), which will assist in performing their daily tasks efficiently. In this diagram, the plant enterprise data, information, and knowledge are represented within database (DB), knowledgebase (KB), and object-base (OB), as well as data files (i.e. ASCII, Image, Video, Sound, etc.).

5.3. CAPE-ModE Design Specifications

In this section, detailed specifications for the different components and layers of CAPE-ModE will be presented. The main components of CAPE-ModE can be summarized as: Model Repository in UML, Model Management System (MMS), Modeling User Interface (MUI) using Model Manipulation Language (MML), Model Query Language (MQL), Model Translator (MT), and Application Programming Interface routines (APIs).

5.3.1. Concept of Model Repository

The cornerstone of the proposed CAPE-ModE is to represent the plant object-oriented model (POOM) in a formal and unified manner using unified modeling language in a central repository. This includes both the technical engineering views and the business enterprise views. UML is developed by OMG, as one of the internationally recognized standard modeling language, to represent the system models prior to the system development using object-oriented approach. Efforts are exerted to develop a formal definition of UML to enable the formal representation of the different plant model elements within the central repository.

The proposed model repository is composed of model base and metamodel base (Modelbase – MB / Metamodelbase – MMB). Example of the metamodel element from the plant design could be a class that represents the behavior of fluids (Gabbar [a], 2000), while pump class could be an example of the model element. The pump class can be associated with the metamodel of the fluid behavior in the form of stereotype (Gabbar [a], 2000). The automated modeling environment will establish such link and offer the design, simulation, and operational system a robust consolidated view of the plant lifecycle model elements.

5.3.2. Model Management System (MMS)

The idea of managing the model elements within a central repository can be fulfilled using a model management system (MMS), which is similar to managing a database using a database management system. The model (i.e. POOM) is composed of model elements that can be described using the metamodel layer. Both metamodel and model layers can be viewed as group of interrelated model elements. The MMS provides similar functionalities to that of the RDBMS (i.e. security & accessibility, retrieval, sorting & searching, etc.). In this context, OO CASE has been used to perform the required MMS functionalities. Objecteering™ has been adopted for that purpose where plant metamodel is constructed using UML profile builder, while the plant model is constructed using the UML builder. MMS will manipulate the different model elements in the form of a formal unified modeling language (UML) within the central repository. MMS offers a set of functions that manipulates the model, such as: Create Model, Create Model Element, Delete Model, Delete Model Element, Change Model Owner, Grant Access Permit, etc.

5.3.3. Modeling User Interface (MUI)

Java™ is used to develop the modeling user interface so that users can manipulate and browse the model (or model elements) via normal Internet browser. Alternatively, MMS can be used to directly access the model elements within the model repository. MUI has mainly three windows: W1, which accepts the input user instructions (MML / MQL),

W2, which displays the results of the execution of the user instructions and commands, and W3, which displays the models and model elements in visual form so that user can select interactively from this list to perform any required model manipulation. MUI includes an intelligent parser that provides online syntax options for both the syntax of the instructions and the parameters retrieved from the central model repository. For example, while user is typing "Gr" system shows a list of instructions that start with "Gr" i.e. "Grant_Access" (based on the user profile and domain). Once user selects "Grant_Access", system will type "Model Name=", followed by a list of existing model names available in the model repository (based on the user permissions and security levels). Sample functions offered by MUI can be viewed in table 5-2.

Table 5-2: Functions Offered by MUI

User Instructions	Accept user instructions (MML / MQL) User instructions syntax checking (Parsing)
Results	List of model elements for a given selection
Model Visualization	Visualize list of models, metamodel elements, and model elements in a visualized form, which also reflect the inheritance among classes

5.3.4. Model Translator (MT)

There are many tools and standards that interact with the plant enterprise models within PEEE. Hence, forcing one standard for model representation may not be a feasible solution. The proposed idea in this context is to enable model translation into different, and common, model representation formats that are commonly used. Examples of such model representation formats are: XML, XMI, Express-X, and HTML. The model is basically represented within the plant model repository (MB/MMB) in a neutral and unified format i.e. UML. The model translator's (MT) main function is to translate the model from/to UML to/from other predefined user formats. This may not be an easy task, however, example has been implemented to convert from UML to HTML and XMI. This automatic conversion is made using an option in the adopted OO CASE tool. Table 5-3 shows a standard XMI code, while table 5-4 shows a sample code for reactor class of HDS plant represented in XMI. The HTML layout of the model can be viewed as in Figure 5-4. In this figure, the HDS process design is divided into control groups called CGU (Naka, 1999). Each CGU is mapped into a model element called packages, which is a logical entity that contains all smaller model elements (Bayer, 1999 and Gabbar, 2000a). Model users can simply click in the hyperlink to navigate through the plant model.

Table 5-3: XMI Standard Code Structure

```
<!-- Document Prologue, etc.   -->
<Model xmi.id="a1"> <name>Business</name><visibility xmi.value="public"/>
    <ownedElement>
        <Class xmi. id="a7"><name>Customer</name>
            <feature>
                <Attribute><name>id</name>
                        <multiplicity><XMI.field>1</ XMI.field>
                                        < XMI.field>1</ XMI.field></multiplicity>
                    <type>< DataType href="|a247"/></type> <!-- Custid -->
                </Attribute>
                <Operation><name>update</name>
                    <scope xmi.value="instance"/>
                </Operation>
            </feature>
        </Class>
    </ownedElement>
</Model>
```

Table 5-4: Sample XMI code for Reactor Class within HDS Plant

```
<XMI.content>
<Foundation.Core.Class xmi.uuid = '655360292:28'>
<Foundation.Core.ModelElement.name>mc-
Reactor</Foundation.Core.ModelElement.name>

<Foundation.Core.ModelElement.visibility xmi.value = 'publi
```

```
<Foundation.Core.GeneralizableElement.isRoot xmi.value = 'fa
<Foundation.Core.GeneralizableElement.isLeaf xmi.value = 'fa
<Foundation.Core.GeneralizableElement.isAbstract xmi.value =
'false'/>
<Foundation.Core.Class.isActive xmi.value = 'false'/>
```

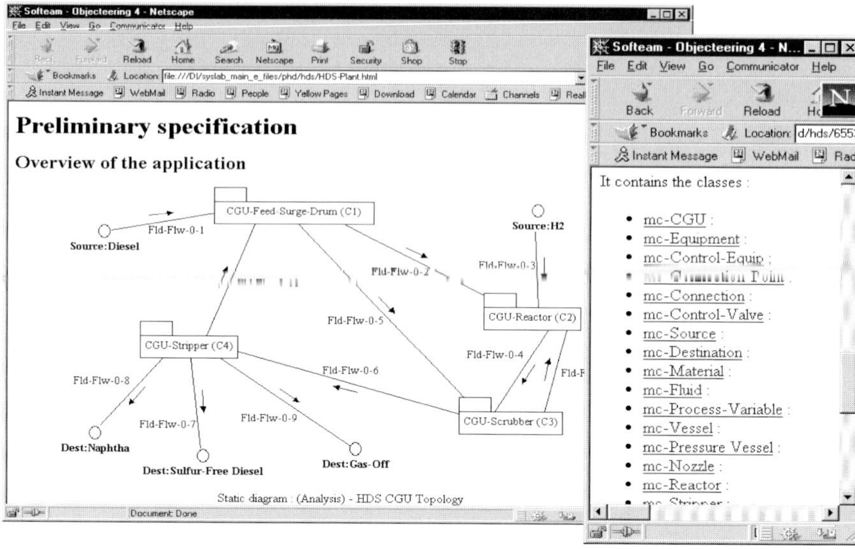

Figure 5-4: Model Representation in HTML Format

5.3.5. API Handler

APIs are used in different ways in the information systems. However APIs are used commonly by external applications / tools to enable accessing the data/knowledge within the domain of one application. In the proposed design of CAPE-ModE, APIs are employed to access the model elements from outside CAPE-ModE domain. Applications and tools will use standard predefined (configurable) APIs to manipulate and query the model elements within the model repository. This can be done via using embedded MML and MQL within the source code of the host application / tool. Table 5-5 shows definitions of examples from the proposed APIs. APIs are configurable routines that enable user customization in the run time. This means that the same API can react differently based on the environment and user settings. The API handler is considered as a server to manage and execute the different APIs from the different applications and tools within the plant enterprise engineering environment. It is considered as a part of the middleware structure of the plant enterprise environment. Hence, using CORBA/IDL could be suitable to build such server program.

Table 5-5: Examples of APIs to Manipulate the Model within CAPE-ModE

Execute MML command	MML instruction to be embedded within the host application / tool to manipulate the underline model.
Execute MQL command	MQL instruction to be embedded within the host application / tool to execute queries about the underline model.

5.4. Model Representation within CAPE-ModE

Once talking about object-oriented approach, it is essential to highlight what are the recent achievements made by OMG. Table 5-6 shows the metadata architecture proposed by OMG, which reflects the four levels of metamodeling (Gabbar, 2000c). Common Warehouse Metamodel (CWM) is a specification that describes metadata interchange among data warehousing, business intelligence, knowledge management and portal technologies. When talking about CAPE-ModE, the plant enterprise models represented in UML can be interchanged among the different domains in CWM format.

Table 5-6: OMG Metadata Architecture

Meta-level	MOF* terms	Examples
M3	Meta-metamodel	The "MOF Model"
M2	Metamodel, meta-metadata	**UML Metamodel**, CWM Metamodel
M1	Model, metadata	**UML models**, CWM Metadata
M0	Object, data	Modeled systems, Warehouse data

**Meta-Object Facility*

Using the proposed CAPE-ModE explained above, the plant enterprise models are presented in UML as a neutral formal language within the model repository. MML is used to manipulate the model elements from the modeling user interface (or through MMS). Similarly, MQL is used as a model query language either from the modeling user interface or through MMS. In this section, these three modeling formal languages will be explained in more details.

5.4.1. UML Formal Definition Initiatives

The informal definition of UML, describes the syntax and semantics of UML, which is used to build and describe the plant enterprise models using object-oriented approach. One of the major outcomes of this study is to draw the outlines of the formal definition of the UML to formally represent the plant enterprise model. Compared with C++ or Express-X, UML is much simpler and more convenient to describe the plant enterprise model using object-oriented approach. Using UML, the plant design, simulation, and operation models can be represented (Gabbar, 2000a). Its' strength comes from objects/classes representation capability using the main three dimensions: static, dynamic, and function. Sample of UML formal syntax can be found in table 5-7. For example, model-visibility can be: "Public", "Protected", "Private", or "Undefined".

Examples of the formal UML definitions have been described by Gabbar (2001c). For the purpose of representing the plant design and simulation models, mathematical equations solver is needed (Linninger et al., 1999). OCL as a standard object-constraint language has been used within UML to enable defining constraints associated with each model element. For that it is proposed to enable the mathematical equations representation within OCL.

Table 5-7: Examples from UML Formal Definition

Model / Model Element	MODEL NAME=<model-name>, OWNER=<model-owner>, VISIBILITY=<model-visibility>
Qualifiers	List of qualifiers can be associated with the model elements, e.g. LEVEL=”abstract” (or “leaf”, Root”)
Tagged Value	Tagged value can be associated with the model element as: Tagged-value-name:Qualifier(p1,p2,p3)
Attribute	Attribute Name=<attribute-name>, OWNER=<attribute-owner>, VISIBILITY=<attribute-visibility>, INTENSION=(<modeling-phase>, <Intension-Description>, <Intension-owner>)
Constraints	OCL is used to formally describe any constraints associated with any model element. Differential equations can be defined within the OCL to describe the behavior of a certain equipment i.e. class [1,2].

5.4.2. Model Manipulation Language (MML)

The main objective of this language is to enable enterprise tools, applications, and humans to manipulate the model elements within the model repository. Table 5-8 shows list of sample MML instructions/commands that can be executed either through the MUI or directly via MMS.

Table 5-8: Example MML Commands

Create Model	CREATE MODEL NAME="HDS-Plant", OWNER="Plant_Modeler", VISIBILITY="Public"
Create Model Element	CREATE MODELELEMENT NAME="Pump", PARENT_CLASS="mc_Pump", OWNER="Plant_Modeler", VISIBILITY="Private"
Delete Model	DELETE MODEL NAME="HDS-Plant", LOG="On", Log-File="c:/model/log-file.dat"
Delete Model Element	DELETE MODELELEMENT NAME="Pump", Log="Off"
Change Model Owner	UPDATE MODEL NAME="HDS-Plant", Owner="Designer", Log="Off"
Grant Access Permit	GRANT_ACCESS MODEL NAME="HDS-Plant", User="Maintenance-User", Access-Type="Read"

5.4.3. Model Query Language (MQL)

Similar to MML, MQL is used to query the model elements using the modeling user interface. Both MML and MQL are understood and executed by the MMS. The result set of the query performed is sent back to the modeling user interface. Both MQL and MML can be embedded in the source code of tools and applications within PEEE. In such case, the result set is sent back to the host application (or tool). Table 5-9 presents example commands from the MQL.

Table 5-9: Example MQL Commands

Retrieve model / model elements	Retrieve model elements (classes, metaclasses) and display their details
Retrieval condition	List of model elements satisfying a given condition

5.5. Mechanism

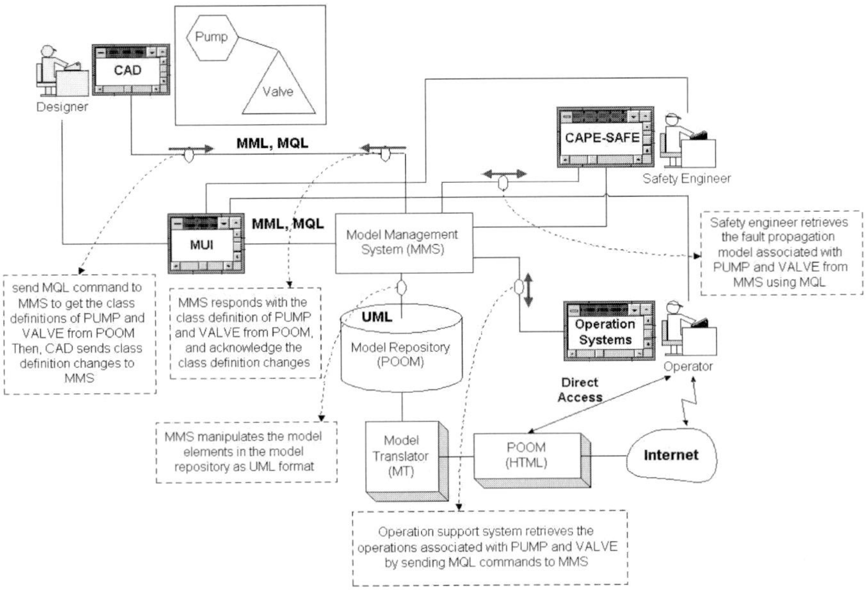

Figure 5-5: CAPE-ModE Mechanism within PEEE

The plant lifecycle starts with the design of the process block diagram. This followed by detailed process design, which includes full engineering details represented in piping and instrumentation diagram or P&ID. For example, within the computer-aided design (or CAD) environment, process designer will select pump class and put in his workspace within CAD environment. Then he selects another physical equipment like valve and connects it to the pump. These actions will automatically create a call from CAD environment to MMS to verify the class definitions of pump and valve within POOM. In case of new class within CAD, MMS creates the new class definitions within POOM. Similarly from plant enterprise safety management environment CAPE-SAFE (Gabbar [e], 2001), fault propagation model can be retrieved from

POOM by issuing MQL commands to MMS. Figure 5-5 shows the interaction between the different components of CAPE-ModE and other components within PEEE. The utilization of such automated modeling environment within PEEE will improve the business functions by providing online and consolidated model elements for the different views.

5.6. Prototype CAPE-ModE

In order to draw a pictorial imagination on the mechanism of the proposed CAPE-ModE within plants/manufacturing enterprise engineering environment, a prototype system has been developed. Figure 5-6 shows the layout of the user interface that enables plant enterprise users, each from his view, to manipulate the different plant model elements. In the top of the main window, user can select the suitable model view by simply clicking the corresponding button. For example, when user selects "DESIGN" button, system automatically brings all model elements that are related to the design models, such as the model elements within the P&ID. Then system displays all the three model views i.e. static, operation, and behavior models, as shown in that figure. User (or modeler) can import or export the model elements in different formats like HTML, XMI, XML, EXPRESS, TEXT, etc. The input window is used to capture user commands in any of the proposed model languages i.e. MQL or MML, while the output window is used to display to the user the results or system messages.

The different contents within the static, operation, and behavior windows can be selected and activated (similar to hyperlinks within the web pages) to display the corresponding model representation in popup windows. For example once user selects "State-1" from the behavior window (shown in figure 5-6) and presses the UML button, system displays a popup window that includes the corresponding state diagram in UML format along with the formal description of the model elements.

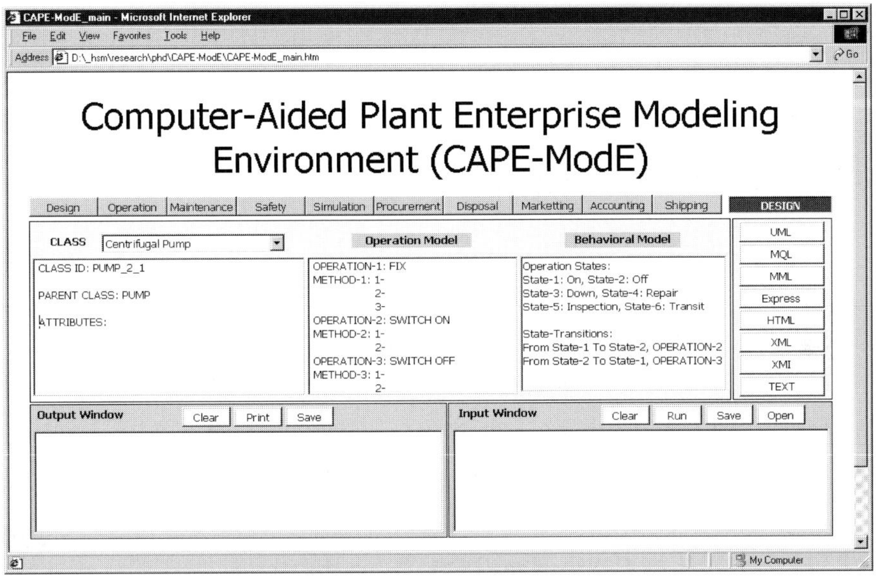

Figure 5-6: Layout of CAPE-ModE Prototype System

Part III: Computer-Aided Plant Enterprise Safety Management System (CAPE-SAFE)

6. Analysis of CAPE-SAFE

Towards the complete development of any computer system, it is essential to carry out the analysis and design phases as part of the software engineering lifecycle. This is a critical success factor to achieve a successful implementation of the underline system. In order to achieve the target goals of CAPE-SAFE, a comprehensive analysis phase has been conducted, as explained in this section.

6.1. Object-Oriented Analysis Methodology

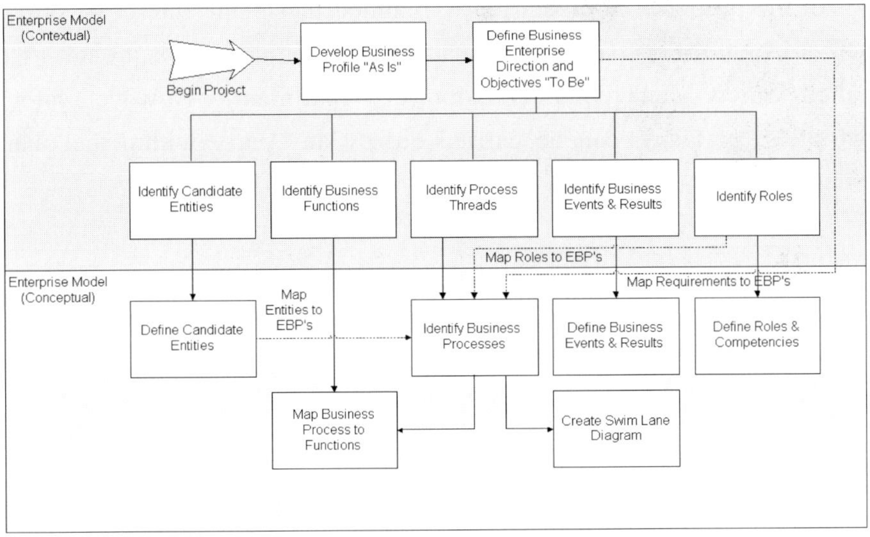

Figure 6-1: Object-Oriented Analysis Methodology

Object-oriented modeling can be divided into two schools: data-centric design and problem-centric design. The problem-centric approach is being used in this work to focus on the problem at hand and move towards the solution. Figure 6-1 shows the overall framework of the modeling approach. The analysis part starts with contextual modeling. This includes developing the business profile "As Is", followed by developing the safety directions and objectives "To Be" for the plant enterprise. This includes identifying business processes that will be decomposed into smaller processes, followed by identifying roles, business events & results, and business functions. The conceptual modeling follows the contextual modeling. It takes the business enterprise model into its detailed level where business functions will be mapped to the business processes. The function decomposition diagram is used to illustrate the function hierarchical model that reflects the relationship between business processes and system functions. The result of this research can be used to complete the detailed logical model of CAPE-SAFE ready for software development.

In the proposed approach, it is assumed that plant enterprise safety activities will be managed by a business area within the plant enterprise called "Safety Management Division (or Department)". However, safety activities (or tasks) can be carried out by any party within the plant enterprise.

6.2. Business Profile "As Is"

The approach starts with the contextual modeling of the "As Is" business model, which describes the current profile of the plant safety division. The profile includes business segments, products/services, customers, suppliers, and other forces impacting the business.

Business segments summarize the key information about the different lines or tracks within the division as in Table 6-1.

Table 6-1: CAPE-SAFE Business Segments

Business Segment	Sub-Segment
Safety Engineering	Hazard Evaluation Hazard Follow-up & Monitoring
Safety Data Management	Accident & Near-miss Data Management Failure & Abnormal Data Management Safety Ontology
Safety Auditing	Safety Regulations Management Safety Procedures Management
Safety Training	Training Scheduling & Management Training Courses & Material

The main products/services offered by the division are:

• Hazard Evaluation Results • Accident / Near-miss Database • Safety Ontology	• Safety Procedures • Safety Regulation • Safety Training

The main customers of the division are:

• Operator • Process Designer • Maintenance Engineer	• Safety Engineer • Manager • Administration Officer

The main suppliers to the division are:

- Other Plants
- International Safety Standards
- Process Design Environment
- Plant Operational Systems

The main competitors of the division are:

- External Safety Consultants
- External Safety Web Sites

The business forces that affect the division are:

- Profitability
- Environment Constraints
- Owner Constraints

6.3. Business Enterprise Directions "To Be"

The next step in contextual modeling is to examine the business directions for safety division. This section outlines the high-level goals, the business objectives, mission statement, etc. that resulted from the analysis. The information sets the boundary for the targeted automated system.

This mission of the safety division is to manage the plant enterprise safety activities and functions throughout the plant lifecycle.

The following are the critical business issues, which requires special attention while developing the automated solution:

- Utilization of hazard evaluation results
- Compliance with safety national and international regulations
- Development of integrated plant enterprise engineering environment
- Utilization of accident database

- Decide the most suitable HE methodology for each situation

The following are the proposed business vision as explored by safety and process management groups:

- Plant safety will be completely managed using CAPE-SAFE, which is integrated with plant enterprise systems and components.
- Plant personnel will be trained (using CAPE-SAFE) on safety engineering and will be aware of the process design model.
- Safety regulations will be fully met and satisfied. The plant will be ahead in the market competition using its excellent.

The following are the business objectives of the safety management division:

- Comply with international safety regulations
- Improve safety awareness among plant personnel
- Conduct and manage hazard evaluation throughout the plant lifecycle
- Integrate Plant Enterprise Safety Management System within PEEE
- Integrate plant safety management system within PEEE

The following are the critical factors to achieve the stated goals and objectives:

- Apply safety design
- Utilize safety information in operational systems
- Utilize other plants safety information
- Develop CAPE-SAFE to manage plant enterprise safety
- Integrate CAPE-SAFE with CAPE-PEEE
- Embed safety aspects within the plant process model

The following are the specific business strategies to achieve:

- To achieve such goal, the plant design model need to be constructed as the core of the plant enterprise systems. Safety model will be constructed and implemented as integrated with the plant model. Plant lifecycle models are also integrated within the same pool using unified modelling language in an online modelling environment that can communicate and interact with systems within the plant enterprise.
- A central safety management is adopted to manage all safety activities within the plant life cycle.
- The safety regulations and procedures will be automated and embedded within the plant model that to be realized in the plant enterprise systems to ensure full compliance with plant safety regulations. Hazard evaluation techniques are automated and integrated with decision support to select and apply the most suitable hazard evaluation technique. The results of the hazard evaluation technique will be monitored and followed-up using an intelligent module. Accident database will be constructed to

provide the proper means to learn from history. Safety ontology is required to resolve any terminology issues.

- Integration with design, maintenance, and operation environment is required to share safety data and to provide any required data into safety domain.

6.4. Requirements Analysis

The enterprise business direction were considered and refined into a set of business requirements. The business requirements must be fulfilled by the business processes and data that the processes deal with.

The following are sample of the business requirements that resulted from the consideration:

- Automate the hazard evaluation process
- Automatically decide the suitable hazard evaluation method based on the selection criteria
- Define all safety terminologies within safety ontology server
- Analyze the safety regulations so that it can be searched, inquired, and/or reported
- Select and adopt safety international standard regulations
- Synthesize safety procedures from the plant safety model
- Build and manage accident database that includes accident data from the current plant and other plants using plant service providers via Internet
- Integrate CAPE-SAFE with plant design environment

- Integrate CAPE-SAFE with plant operational systems
- Monitor & follow-up hazard evaluation results
- Utilize computer-based training and web-based training for safety training
- Identify the stages where hazard evaluation is required
- Monitor hazard causes throughout the plant lifecycle

6.5. Safety Solution Challenges

The following table summarizes the challenges that faces the industries to implement integrated safety solutions:

Table 6-2: Integrated Safety Solution Challenges

Issue	Challenges	Proposed Solution
Safety Training	Need to be updated with the online company data	The safety training module need to be integrated within plant enterprise engineering environment and to have access to sample plant design and operation actual and simulation data.
	Need to be continuous, and with monitoring facility	Web-based and computer-based training can help new employees to practice the safety training courses, and also can help management to monitor employees performance with respect to the their safety performance.

Table 6-2: Integrated Safety Solution Challenges (Cont.)

Issue	Challenges	Proposed Solution
Safety Training (Cont.)	Need to reflect the latest national and international safety regulations and legislations	Safety training module will be integrated with safety regulations manager that maintains the latest safety regulations and legislations.
	High training cost	Achieve the minimum training cost with the use of web-based and computer-based training.
Safety Regulations	Safety regulations are manually developed and maintained	Use of knowledge engineering concepts to structure and automate the development of the safety regulations. This will facilitate the automatic update from the source of the regulations (i.e. national or international safety regulation authority) to all clients (plants or engineering companies).
	Safety regulations are manually (hardcopy) transferred from source to destination	The use of plant service provider who can automatically download / upload the latest regulations using standard structure format.
Safety Procedures	Safety procedures are developed and maintained manually	Automatic generation of the safety procedures using the plant model.
Utilization of existing hazard evaluation results	There are many hazard evaluation results that are available, which include useful information for plant safety practices. It is essential to find a systematic way to utilize these information	Development of a systematic mechanism to extract fault propagation results (i.e. cause/consequence) from the existing hazard evaluation results. Propose fault propagation model that can represent different fault schemes. Design software program that can extract the fault propagation model and hazard evaluation results from existing text-based hazard evaluation results.
Safety Model	No existing safety model that can abstract safety aspects within the plant lifecycle	The absence of complete conceptual safety model that can represent the different safety aspects within the plant lifecycle made it difficult to develop more realistic safety solution. For that, a conceptual plant enterprise

		safety model has been developed to identify the building blocks of complete plant enterprise safety model.
Safety Design	No clear method to link safety aspects (enterprise safety model) to process design.	The attempts started with development business activity models that represent the safety design activities. This has identified the integration points between the safety and design domains. However, by utilizing unified modeling language to represent the plant design model enabled us to associate all safety aspects with the different plant model elements (static, behavior, and operation).
Integrated Safety Solution	Till to date, no complete and integrated safety solution that can cover all safety functions throughout the plant lifecycle.	The design of the proposed enterprise safety solution will enable industries to automate all safety functions throughout the plant lifecycle. The proposed solution covers all functions that are highlighted during the analysis phase.

6.6. Process Threads

Within safety management division, there are major process threads that perform the major safety functions. Table 6-3 shows the major process threads within plant enterprise safety:

Table 6-3: Process Threads

Name	Initial Event	Primary Result
Build Accident Database	"Accident Occurred" "Incident Occurred" "Accident Records Downloaded from Internet"	"Accident Report" "Online Acknowledgement of Similar Accidents to Designer" "Online Acknowledgement of Similar Accidents to Operator"
Collect Safety-Related Data from Maintenance	"Safety-related Maintenance Task is Completed"	"Safety-related Data From Maintenance System"
Collect Safety-Related Data from Operational Systems	"Operational Task is Completed"	"Safety-Related Data from Operational Systems"
Conduct Hazard Evaluation	"Process Design of SS is Completed" "Process Design Phase is Completed" "Process Design is Changed"	"Severity of Each Propagation Path" "Frequency of Initial Event" "Probability of Each Propagation Path" "Risk Level of Each Propagation Path" "Propagation Speed of Each Propagation Path" "Hazard Evaluation

		Results"
Decide Suitable Hazard Evaluation Method	"Request to Conduct Hazard Evaluation"	"Selected Hazard Evaluation Method"
Follow-up Hazard Evaluation Results	"Hazard Evaluation Results Available"	"Operational System Updates" "Maintenance System Updates" "Safety Management Report"
Manage Safety Regulations	"New Safety Regulations is Adopted" "Safety Regulations Materials are Accessed"	"Safety Regulations Model" "Safety Regulations Database" "Safety Regulations Script"

Table 6-3: Process Threads (Cont.)

Name	Initial Event	Primary Result
Design Trigger for Hazard Evaluation	"Process Design of S/U & S/D is Completed" "Process Design of Abnormal Situation is Completed (IPL3,4,5)" "Process Design of SS is Completed" "Process Design of E/S is Completed" "Process Design of Emergency Shutdown is Completed"	"Hazard Evaluation Triggers"
Manage Safety Ontology	"New List of Safety Terminology is Downloaded From Internet"	"New List of Safety Terminology is Developed" "List of Equivalent Safety Terminology" "List of Not-Used Safety Terminology"
Manage Safety Training	"New Employee is Hired" "Process Design is Changed" "Start Plant Operation" "New Safety Regulations is Introduced"	"Safety Training Materials" "Training Schedules for Plant Personal"
Process Design of Abnormal Situation is Completed (IPL3, 4, 5)	"Process Design of Recovery (Fallback) is Completed" "Process Design of Partial Shutdown is Completed" "Process Design of Total Shutdown is Completed"	"Functional Equipment Specifications" Soft Sensors Information" "Flowsheets Information" "Time and Cost Estimation"

6.7. Business Process Chart Diagrams

The process chart diagram is a process flow diagram for modeling business events, processes, and results. Figure 6-2 shows the business process chart diagram that represents the conduction of hazard evaluation. In this diagram there are two swimlanes: Process design unit, which represents the process design unit; and safety, which is the plant enterprise safety management division. There are mainly two initial events that initiate the hazard evaluation process to be executed: Process design is changed, and/or process design phase is completed. The initial event of "Process Design is Completed" is expanded further as in figure 6-3.

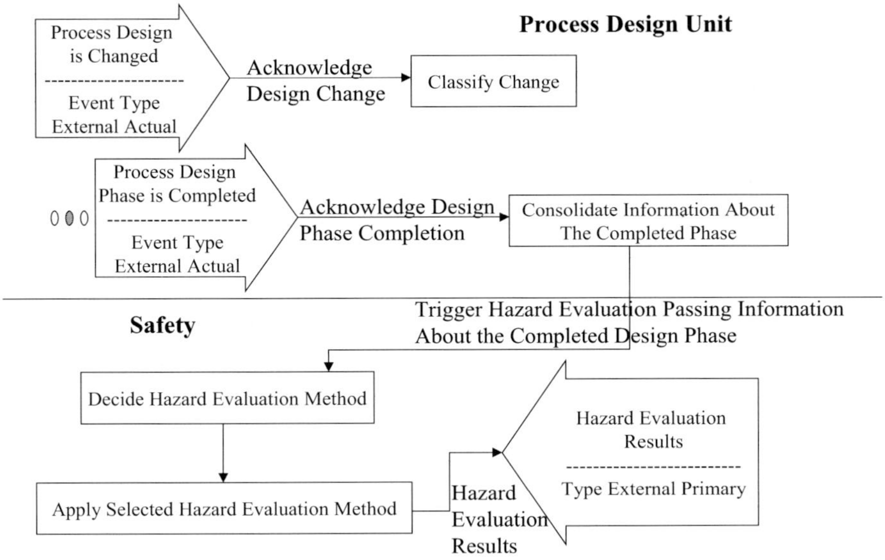

Figure 6-2: Hazard Evaluation Process Chart

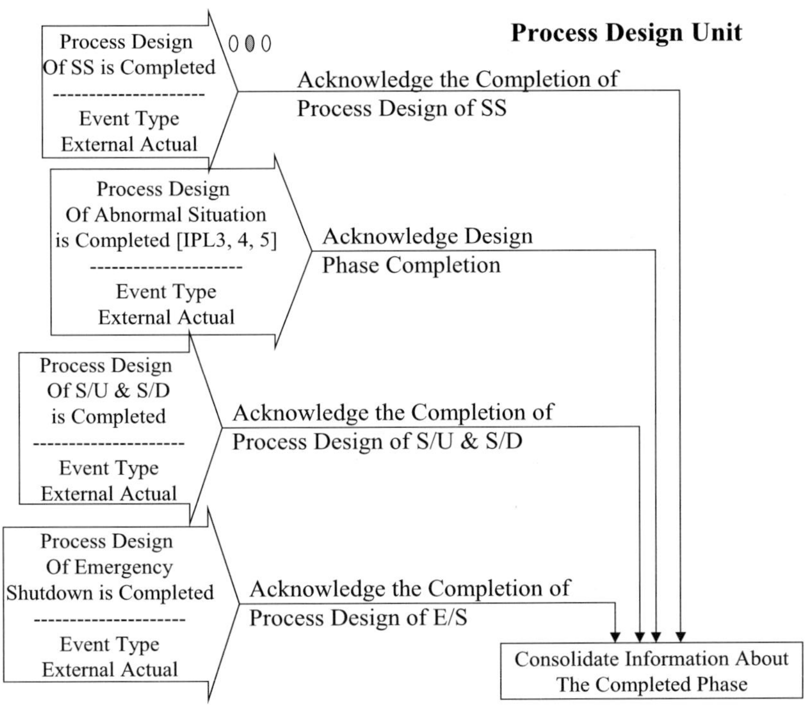

Figure 6-3: Design Completion Process Chart

6.8. Safety Design

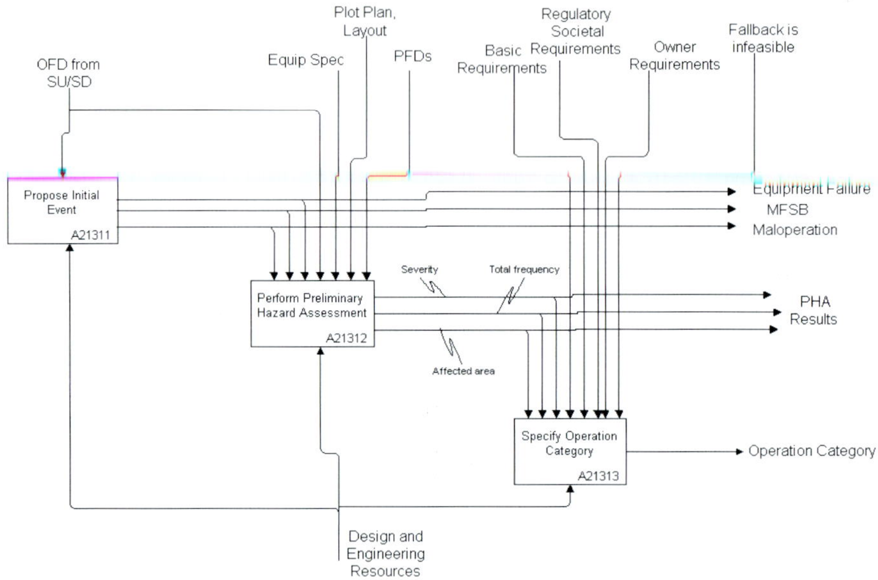

Figure 6-4: Example of Design Activity Model

As part of the analysis phase, it is essential to consider all safety-related activities that are carried out during the design stage. A research group, composed of Tokyo Institute of Technology, Kyushu University, and Okayama University – Japan, are dedicated to derive the complete business activity modeling of the process design stage, using IDEF0 methodology. The concept of operational design (Naka, 1999) and safety design are adopted to consider plant lifecycle activities while focusing on process design. This includes process safety. The output of this research group is a complete set of business activity model hierarchy. From these activity models, all safety-related activities, inputs, and outputs are highlighted and considered within the CAPE-SAFE domain. Figure 6-4 shows an example of the process design activity models, which includes safety-related activities such as perform preliminary hazard assessment, while severity, affected areas, and total frequency are examples of the outputs that CAPE-SAFE should be able to provide to design environment. The concept of integrating safety environment (CAPE-SAFE) with design environment is useful to centralize all safety-related activities and outputs within CAPE-SAFE, while inputs are provided by the different systems and environments within PEEE to invoke safety-related activity (or function).

7. CAPE-SAFE Design

As a part of the software engineering lifecycle of the development of CAPE-SAFE, user requirements have been conducted on Oil & Gas plant where one of their management objectives is to develop and implement an integrated plant enterprise safety management system. As a result of the analysis phase the main components of CAPE-SAFE have been identified along with the expected functions for each component.

7.1. CAPE-SAFE Components

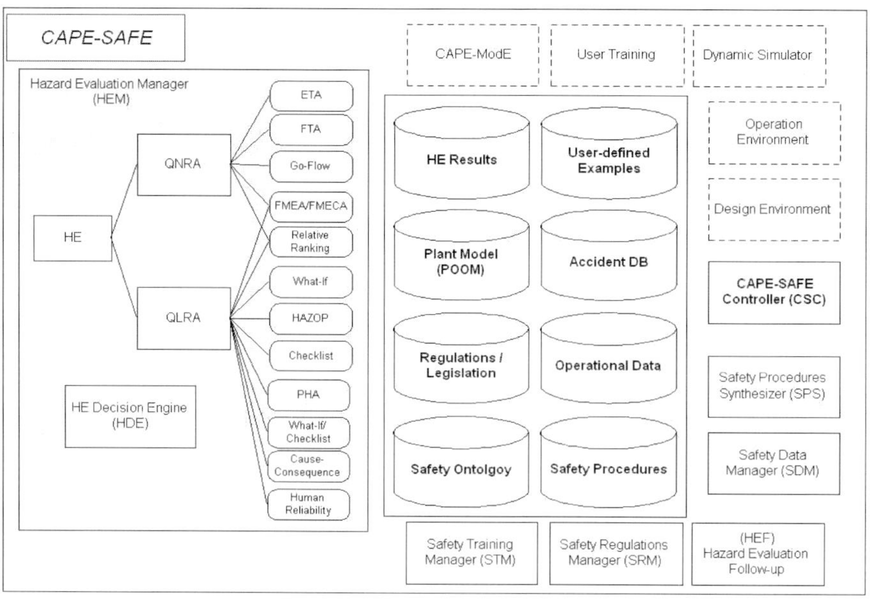

Figure 7-1: CAPE-SAFE System Architecture

CAPE-SAFE is part of the plant enterprise engineering environment that is aimed to manage all safety related activities throughout the plant lifecycle. To achieve that, CAPE-SAFE is designed as an integrated environment as in figure 7-1.

In this section, detailed module specifications will be presented.

7.1.1. Hazard Evaluation Manager (HEM)

HAZOP is one of the commonly methods for hazard and operability analysis. However, it may not be the best for all hazard evaluation cases or situations throughout the plant lifecycle. Hence, we are proposing an integrated hazard evaluation manager (HEM) module that can apply the most suitable qualitative and/or quantitative hazard evaluation method for each situation based on selection criteria. A built-in decision engine (expert system) is used to decide intelligently the most suitable hazard evaluation method based on the selection criteria.

7.1.2. Hazard Evaluation Follow-up (HEF)

HEF is dedicated to monitor and follow-up the hazards that occurred at any stage within the plant lifecycle. Specialized team members can acquire insights of the hazards that exist in the underline process and make suggestions that might help reducing the risk of hazardous situations. An intelligent system (expert system) can assist in this process by providing similar results with more suitable suggestions that match the underline process. This will help in controlling discovered hazards, avoiding similar hazards to occur, and ensuring that the corrective action has been taken. Using HEF, plant can achieve higher quality, incur lower operating costs, and provide managers greater confidence that hazards have been identified and are being controlled (CCPS).

7.1.3. Safety Data Manager (SDM)

This module is responsible to capture all safety data from all other modules within PEEE (through interfaces) and update the safety database and knowledgebase. Example of one data group that is maintained by SDM is the accident/incident data from the current plant as well as from other plants (via Internet – PSP). Safety ontology is another data group that is constructed and maintained to provide plant personnel with the right safety terminology and standard terms and definitions. This data group can be used to resolve issues and conflict related to safety terminologies.

7.1.4. Safety Procedures Synthesizer (SPS)

As safety procedures are as essential as the safety assessment practices, and since the safety procedures are can be changed based on the plant design and operating condition, hence, it is essential to automatically generate the plant enterprise safety procedures. SPS module will automatically generate the safety procedures throughout the plant lifecycle. This requires input from plant modeling, design, and operational environments.

7.1.5. Safety Regulations Manager (SRM)

In order to manage and apply safety regulations, it is essential to analyze national, international, and corporate safety regulations and incorporate them into the plant enterprise safety environment in a structured manner so that it can be checked, applied, managed, and/or monitored. This requires online update from external sources (i.e. international standard web sites) as well as amendment facility by the enterprise safety personnel. SRM module will provide such facilities as integrated within CAPE-SAFE.

7.1.6. CAPE-SAFE Controller (CSC)

The different modules within CAPE-SAFE will be controlled using CSC module. This module will control the information flow among the different modules within CAPE-SAFE and also provide control, validation, and security on information and instructions flow with other components within PEEE.

7.1.7. Safety Training Manager (STM)

Most of the plant safety engineers have reported that safety training is an essential part in the plant enterprise safety. This is because the most effective way to improve plant safety (or reduce the risk) is by increasing the safety culture and education among plant personnel. This can be achieved through continuous training in any form (i.e. computer-based training, on the job training). STM module will be used to manage all safety-related training activities.

7.1.8. Safety Data/Knowledge Groups

CAPE-SAFE will be the owner of safety-related data and knowledge groups. The following are some examples of the safety-related data and knowledge groups that will be maintained and managed within CAPE-SAFE: Hazard Evaluation Results, User-defined Scenarios and Examples for Hazard (or Abnormal Situations), Safety Procedures, Safety Regulations / Legislations, Safety Ontology, Operational Data, Accidents/Incidents/Near-miss, and Plant OO Model (POOM).

7.2. CAPE-SAFE Integration in PEEE

The complete picture of CAPE-SAFE within PEEE, as shown in figure 4-5, indicates that CAPE-SAFE has links with design, operation, enterprise modeling and plant enterprise environments.

7.2.1. Integration with Modeling Environment

Plant enterprise safety model has been proposed by Gabbar (2000c), where safety criteria are described using object constraint language (OCL) as constraints associated with the plant model elements represented using unified modeling language (UML) from (OMG). This requires an online interface between CAPE-SAFE and the modeling environment to establish the processes depicted from the analysis phase, as shown in table 7-1.

Table 7-1: Examples from the CAPE-SAFE Integration with Plant Modeling

Initial Event	Result	Interface Direction
Equipment-Class is Changed	Validate plant model changes weather it require hazard evaluation process or not.	From: Modeling To: Safety
Safety criteria is changed against control valve classes	Modify the safety constrain associated with the control valve class	From: Safety To: Modeling
Develop new safety constraints	Define and associate safety constraints as associated OCL to model elements	From: Safety To: Modeling

7.2.2. Integration with Design Environment

Any design changes during the design stage, will be validated weather hazard evaluation is required or not. Hence, information about the design changes is passed to CAPE-SAFE for discretion and action. While safety assessment information i.e. hazard evaluation results are highly needed for process designer to debate the different design rationales.

7.2.3. Integration with Simulation Environment

Within the safety evaluation practices, there are many solutions that are based on identifying the initial event, which is used as an input to the dynamic simulation to conduct process simulation to highlight any abnormal situation that may arise. Such abnormal situations can be quantitatively assessed to show the severity level and risk. Such experiment has been made by many researchers as a part of the process safety design i.e. IPL design. Also such information are important for operator to assist in his daily operation.

7.2.4. Integration with Operation Environment

Safety evaluation results are very essential for operator. In fact such data can't be found in any other place within the PEEE. Hence, providing such data can add value to operator in his daily work. HAZOP results can help operator in understanding the cause and consequence of the different failures that may occur to any part of the plant process. Operator can use this data to decide the actual cause of any given error, and hence, can decide the most suitable corrective action. For that hazard evaluation results are shared with the operation environment. From operational systems, the daily failure in any part of the plant equipment can help in updating the failure database that is used in the hazard evaluation process.

7.2.5. Sharing Common Services with PEEE

CAPE-SAFE is one of the automated tools that will be integrated within PEEE. A group of shared services will be shared by all components within PEEE. Examples of such shared services may include: Document Manager, Work Flow Manager, Report Writer, Plant Service Provider, Data Exchanger, Data Warehousing, Graphics Library, and other useful services that can be shared among the different components within PEEE.

7.3. CAPE-SAFE Implementation within PEEE

In this section, the implementation of CAPE-SAFE, as a part of PEEE, will be discussed in more details using a selected case study from chemical / petrochemical plant.

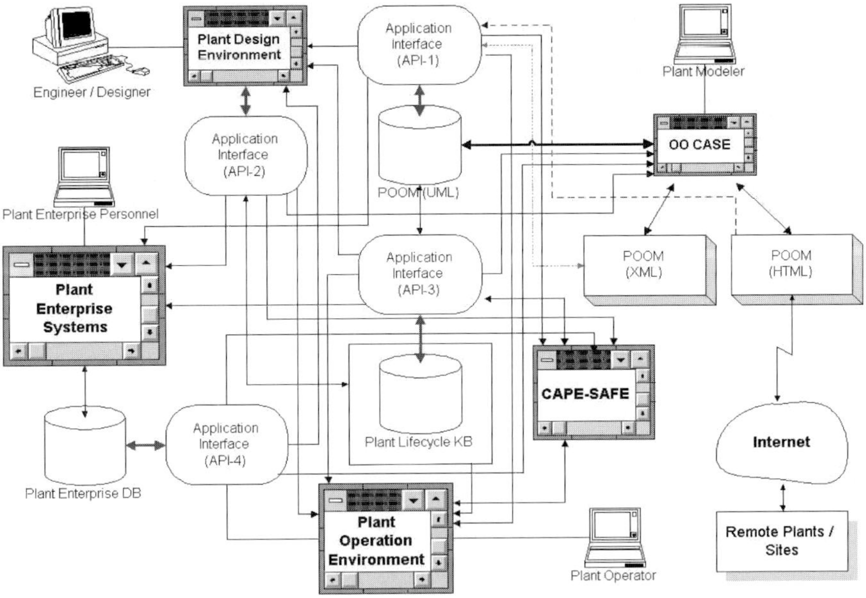

Figure 7-2: CAPE-SAFE Implementation within PEEE

7.3.1. Information System Architecture

Figure 7-2 shows the detailed information system architecture for implementing CAPE-SAFE within PEEE. In this figure, CAPE-SAFE is connected through the different APIs to plant lifecycle knowledgebase & database, plant enterprise applications, and design and operation environments. The plant model (POOM) is the cornerstone of such implementation where safety aspects are manipulated to assist in the different functions that carried out by CAPE-SAFE.

7.3.2. Implementation

In order to implement the proposed CAPE-SAFE solution, it is required that all parties have access to CAPE-SAFE via Internet (on the application server). Clients from plant owners, operating company, or engineering company can manipulate safety-related data/information/knowledge using the different modules of CAPE-SAFE. Java will be used as a programming language where JDBC will be used to establish the connection to the different databases, while APIs will be embedded to manipulate the knowledge within the knowledgebase i.e. G2. Other APIs will be used to connect to the specific data files i.e. XMI/XML/UML representing the plant model and CAD files representing the plant design.

The implementation will commence after the completion of the development of the specified system. In order to be able to capture safety related information, and to manage the safety assessment practices that are usually carried out in the early beginning of the plant lifecycle, CAPE-SAFE need to be implemented and to be running from the concept stage within the plant lifecycle. At that stage, owners usually study the concept of designing and operating the plant prior to plant design stage, which includes some safety constraints and procedures with respect to the plant process block diagrams. The operating company, engineering company, owners, and suppliers will need to interact with CAPE-SAFE to exchange and manipulate safety-related information.

In order to visualize the proposed CAPE-SAFE, a prototype system has been developed. Figure 7-3 shows the proposed implementation system architecture where users are connected to CAPE-SAFE application server using a normal browser where the different applets of CAPE-SAFE will be downloaded into the client's browser. Access to database, knowledgebase, and data files will be carried out using the proposed information system architecture within PEEE.

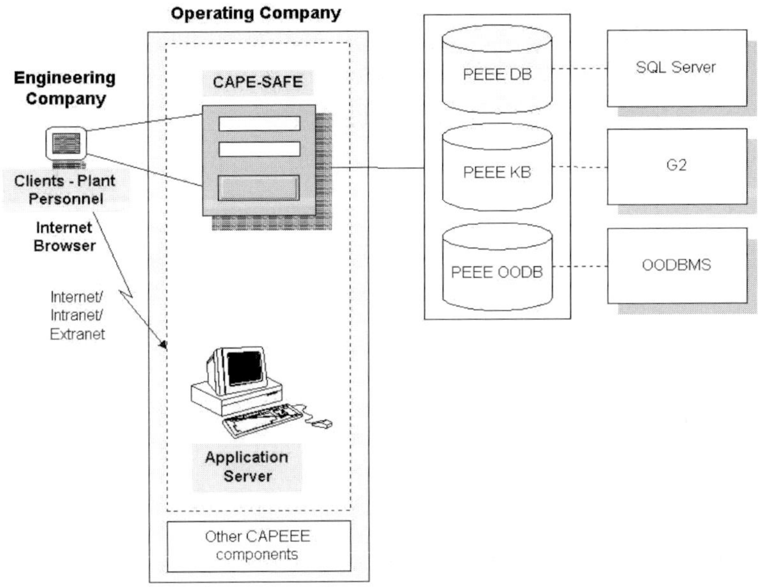

Figure 7-3: CAPE-SAFE Implementation

7.4. CAPE-SAFE Prototype System Development

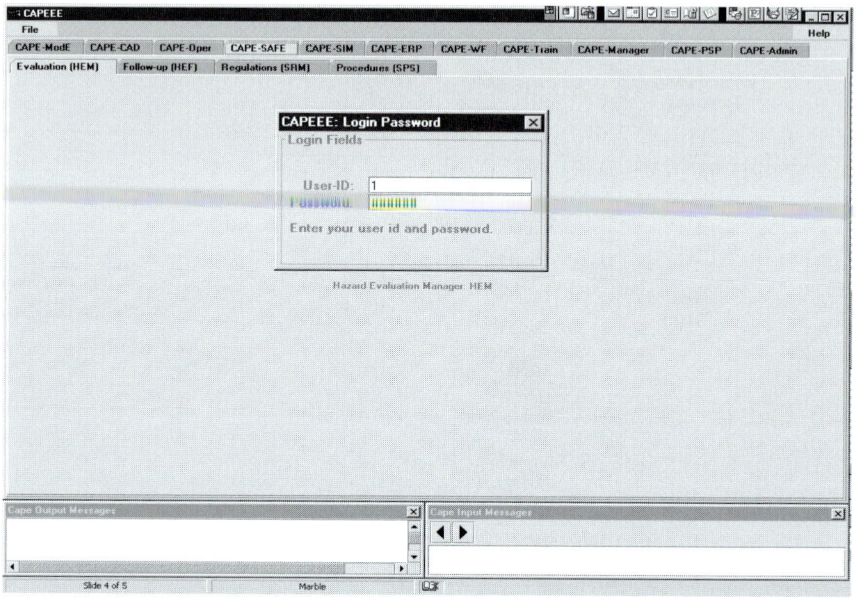

Figure 7-4: Prototype CAPE-SAFE System

Prototype system has been developed using Java; JDK 1.3 from Javasoft; as a programming language, while SQL Server has been used to store the plant enterprise information/knowledge. Screenshots of the developed prototype system are illustrated in figure 7-4. The prototype system has been developed so that it supports the concept of the concurrent engineering where three windows are used by plant users all the time. The main window has been designed as multi-tabs, each tab represents one view of the system i.e. design, operation, safety, etc. Another window is used as input window where users are allowed to input selection criteria or questions. The output window appears in the bottom left of the screen, which is used to display answers and output of user questions and queries. Within each tab, there will multiple sub-tabs where more detailed functions are managed, for example training tab will be subdivided into sub-tabs for safety training, operation training, etc. The benefit of such design is that users can look to the plant lifecycle from different views, for example operator can view the safety training in the training tab, while looking to the plant design in the design tab, and also can perform his daily work in the operation tab. All access rights and permission are subject to the enterprise security model where user groups will be granted special access permission based on their daily work and job description. User ID and password are validated against records in database called "PEEE" in SQL Server" using JDBC connection from the Java program.

7.5. CAPE-SAFE Function Decomposition

As part of the design of the proposed solution, the high-level function decomposition of CAPE-SAFE is shown in figure 7-5.

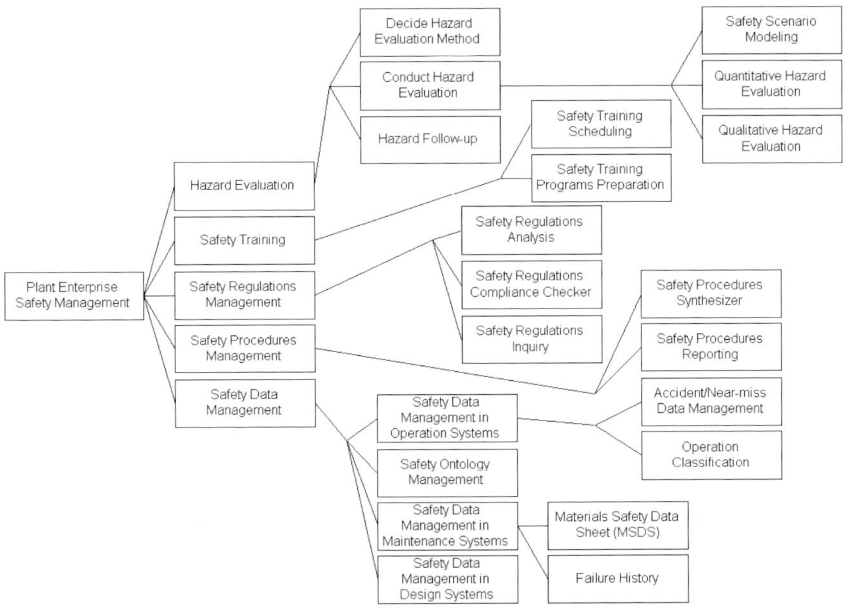

Figure 7-5: CAPE-SAFE Function Decomposition

7.6. Positioning with CAPE-OPEN

CAPE-OPEN (Global CAPE-OPEN) standards have been developed by international research groups centered in Europe to provide a standard and integrated environment for all computer-aided process engineering components. It simply enables the development of any component to be integrated within CAPE suite using the middleware standard interfaces. CAPE-SAFE is intended to be a new component within the CAPE environment to be utilized by any plant to manage lifecycle safety. This will enable the complete integration with other CAPE components. Currently, CAPE-OPEN group is offering subscription and membership to use such CAPE facilities with web-based services management. The ultimate goal is to enable user-defined their own services as compliant with CAPE standards so that others users can use such services.

Part IV: Utilization

8. Mechanism

In order to define the detailed mechanism of CAPE-SAFE it is essential to define the business model of all related functions like design, operation, and maintenance. A team has been established between Tokyo Institute of Technology, Kyushu University, and Okayama University to carry out this task. The result is a complete set of business activity models, which cover the design phase while considering safety and operation factors, under the name of "safety & operational design". The developed business models are used as a reference in this book to define the design business processes that requires interaction with safety environment i.e. CAPE-SAFE.

8.1. Safety Data Management

One key feature in CAPE-SAFE is the management of all safety-related data via the interfaces with other components within PEEE. Safety data groups can be viewed as in table 8-1.

Table 8-1: Safety Data Groups Management Mechanism

Data Group	Data Elements	Management Mechanism
Hazard Evaluation	Hazard, Plant, Process, Equipment, Process Variables, Position, Cause, Consequence, Initial-Event (Root Cause)	- Develop/generate by safety experts/engineers/auditors - Follow-up and monitor by CAPE-SAFE and safety experts/engineers/auditors

Accident / Near-miss	Accident, Plant, Process, Equipment, Process Variables, Position, Cause, Consequence, Date, Location	- Download from PSP - Maintain by CAPE-SAFE from maintenance and operational systems - Inquire by design, operation, and maintenance environments
Safety Procedures	Keyword, Condition, Rule, Action, Owner, Participants, Validation, Plant, Process, Equipment, Position, Operation	- Download from PSP - Generate and maintain by CAPE-SAFE - Assess by safety experts/engineers/auditors
Safety Regulations	Keyword, Condition, Rule, Action, Owner, Participants, Validation, Plant, Process, Equipment, Position, Operation, Assessment Description	- Download from PSP - Develop by plant expert personnel - Assess by safety experts/engineers/auditors
Safety Ontology	Term, Description, Alternatives, Linked Terms	- Download from PSP - Develop by plant expert personnel - Maintain by CAPE-SAFE modules

8.1.1. CAPE-SAFE Conceptual Data Model

Figure 8-1 shows the data groups of the conceptual data model of CAPE-SAFE. In this diagram, PSP index includes the data related to downloading and uploading of safety data with PSPs. For consistency purposes, physical, behavior, and operation models of the plant process are represented in this picture although they are part of the plant design environment.

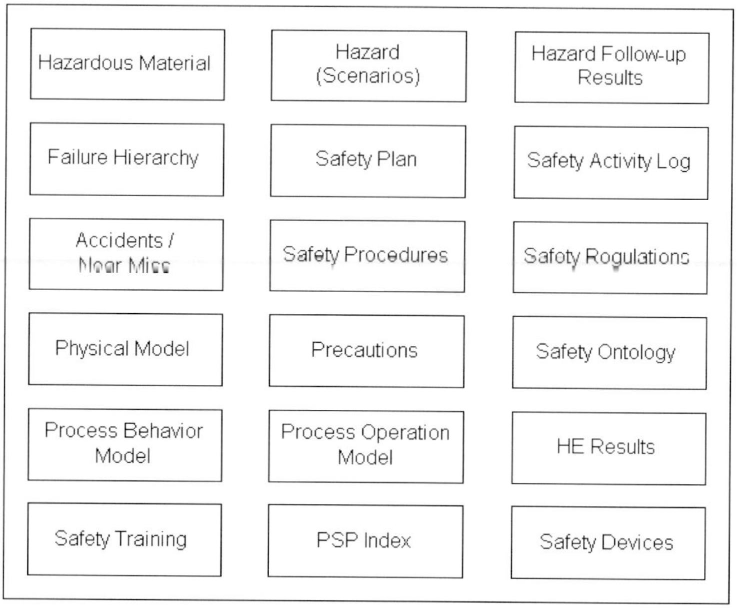

Figure 8-1: CAPE-SAFE Data Model

8.1.2. Safety Historical Data Structure

Safety historical data has been highlighted as part of the conceptual plant safety model. Figure 3-11 showed the detailed data groups of safety historical data component where safety historical data is divided into two parts: plant specifications, which covers all specification changes due to design, operation, or any other process throughout the plant lifecycle; and accident/incident data, which includes plant, events, process, equipment, cause, consequence, and procedures in-place.

8.2. <u>Physical Data Model Specifications</u>

In this section the physical data model of the proposed CAPE-SAFE within PEEE will be specified in more details.

8.2.1. Plant Static Model

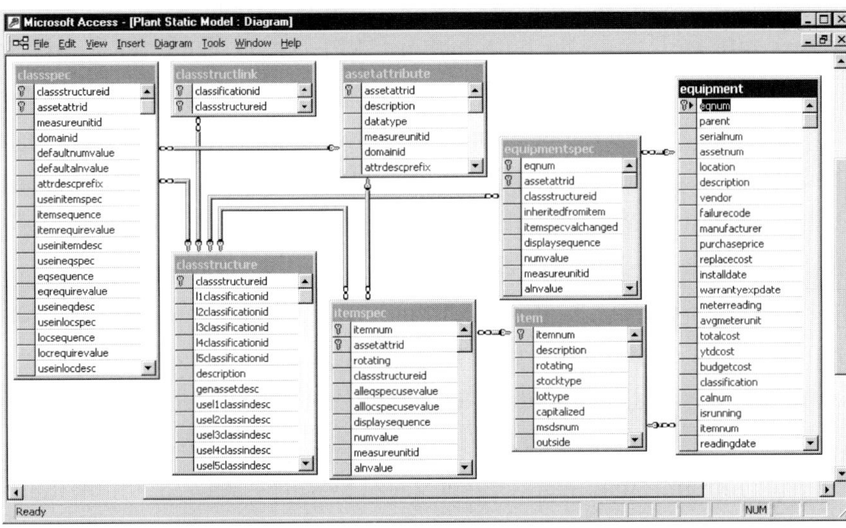

Figure 8-2: Schema Diagram of Plant Static Model

Figure 8-2 shows the relationships between the different entities of the physical data model that represents the plant physical structure. In this figure, plant equipment class specifications are represented in "equipmentspec", while the equipment records are represented in "equipment". Similarly for spare parts, "itemspec" is used to specify the class specifications while "item" is used to describe the spare parts instances i.e. objects. The specifications of the class hierarchy are represented in "classstructure", "classspec", and "assetattribute".

Examples of the data structures for different data groups within CAPE-SAFE are illustrated in appendix 3.

8.3. Automated Hazard Evaluation Results Structuring

One major advantage of CAPE-SAFE is the utilization in automating the structuring of hazard evaluation results produced manually by experts into structured formats within CAPE-SAFE using the proposed fault propagation model (FPM). This has significant importance as till to date there are large number of hazard evaluation practices that have been conducted in chemical/petrochemical plants. Most of these hazard evaluation results are not utilized and kept as history documents. These studies contain significant results that need to be considered by plant lifecycle personnel i.e. designer, maintenance, operator, and safety engineer. The main idea is how to convert these results into electronic format automatically so that it can be integrated with other data groups in CAPE-SAFE within PEEE. The proposed idea is to use intelligent server; program; that can extract the keywords related to the fault propagation model elements and link the cause to the consequences in a structured way to be accessed and manipulated by the different systems and humans interacting with CAPE-SAFE. Figure 8-3 shows the design concept of this idea where hazard evaluation results expected to be at least in text format; if not can be converted using optical character recognition (OCR) programs. The intelligent program will scan line-by-line to extract the cause and consequence of each hazard and store them within CAPE-SAFE. This process will be done automatically using the predefined fault propagation model elements lists.

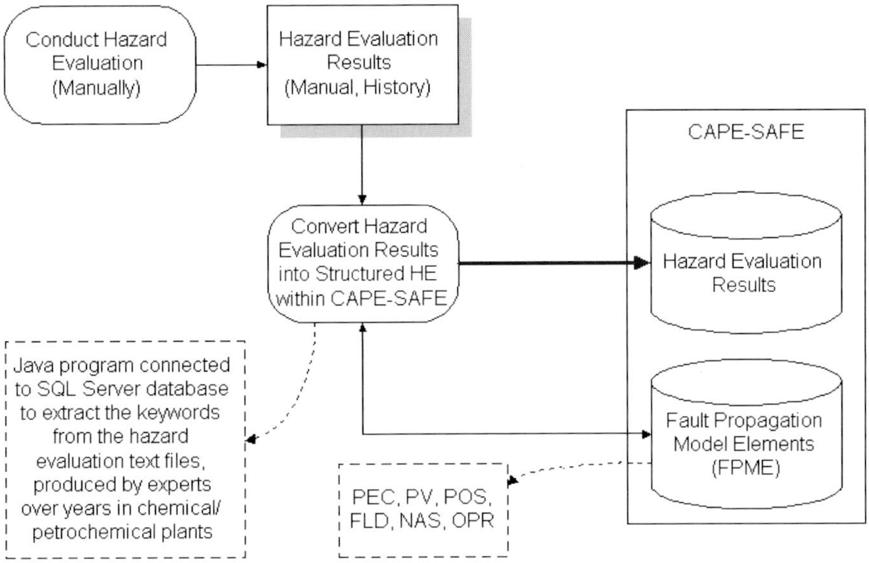

Figure 8-3: Hazard Evaluation Results Automatic Structuring

8.4. Safety Regulations

Safety regulations represent the control and evaluation criteria of the level of risk or plant safety. Since the safety regulations are currently represented in the textual formats, so it is essential to develop a mechanism of how to convert these safety regulations to be active and structured. This requires analysis and formatting of some examples of the safety regulations so that it can be converted into structured and interrelated objects. The first issue in this subject is how to represent the safety regulations so that it can be searched, queried, stored, retrieved, and assessed. This work has been initiated by many researchers in the area of natural language processing and artificial intelligence mainly in the ontology approaches where the underline text is required to be converted to its basics so that it can be converted and manipulated.

These techniques will enable CAPE-SAFE to provide the following facilities i.e. functions:

- Download national or international safety regulations, grouped by subject or keywords
- Understand and assess the modified safety regulations
- Store, retrieve, and search the safety regulations

Further study is required to address the detailed mechanism of this component.

8.5. Safety Procedures

There are many attempts that tried to generate safety or operating procedures. These attempts tried to generate the procedures using artificial intelligence techniques like the work done by Loughborough University – UK. The proposed approach in this book is based on generating the related operating procedures from the plant object-oriented model.

During the modeling practices of the plant lifecycle, procedures are defined as the steps within each operation and as set of constraints associated with the different model elements of the plant lifecycle model. This will facilitate the automatic generation of the related safety procedures associated with the target process within the plant. This has been represented to operator in the form of HTML pages.

8.6. Safety Training

The examples seen from the industry showed that large amount of the budget is allocated for safety training. One company has 1000 technicians/operators/labors requires around 20 hours training for each employee every year to practice all safety related issues to his job. If the estimated cost is US$250 for one hour training (including the cost of leaving the field, course material, etc.), this means that the company requires US$1000 * 20 * 250 = US$5,000,000 to train their technical personnel every year, which is extremely large amount. Also the quality of training material, instructors, and sample data are essential to be considered while conducting safety training. The computer-based training (CBT) and web-based training (WBT) are offering some solutions to such problems. However, the problem of the training material and data has not been solved yet. General safety training courses may not help specific plant to focus on certain aspects of their operation or nature of process. The solution is to have customized safety-training courses for each plant, which is again another cost. The idea presented here about safety training can reduce the cost of training to the minimum. This can be achieved by integrating safety training module with the plant design and operation within PEEE, as part of CAPE-SAFE. This means that the training scenarios are generated while performing plant design, and operation for fault detection. Also process related data are used from the plant design environment, and plant failures and historical data can provide the trainees with real data and problems that occur in the real plant operation. This requires developing intelligent module to extract safety related data and organize them in a shape that is suitable for the training courses, grouped by the training course subjects.

It is expected that the implementation of this module i.e. safety training manager (STM) will improve the safety level of plant operation and personnel, hence reduce the risk and cost of operation.

9. Case Studies

In this section, more details about how the different components and modules within CAPE-SAFE will work within the integrated picture of PEEE. Case studies will be used from different plant types i.e. continuous and batch plants, as well as oil refinery process to show the mechanism of CAPE-SAFE following some scenarios. Other scenarios will be illustrated for the integration with some components within PEEE such as fault detection system, RCM-based CMMS, and operator interface system.

9.1. Examples from HDS Plant

Hydrodesulfurization (HDS) is a continuous plant used to produce Naphtha and Diesel from H2 and raw diesel using heating and reaction chemical processes. Figure 9-1 shows the block diagram of the HDS process, which is composed of 4 main blocks: Feed Surge-Drum, Reactor, Scrubber, and Stripper, while the input is Diesel and H2 and the output is Naphtha, Diesel, and Gass-off.

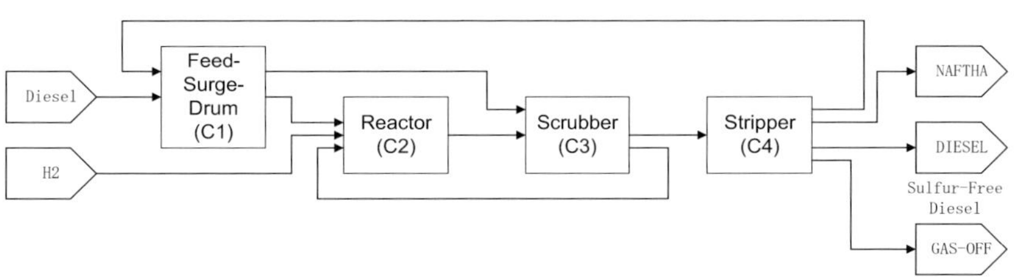

Figure 9-1: HDS Block Diagram

In order to realize safety aspects within the plant model, it is required to represent the plant design model in object-oriented manner within POOM, and then associate safety aspects with the model elements.

Reactor process has been selected in this example. The P&ID of HDS plant has been divided into control group units or CGUs, which is used to analyze process safety in smaller parts of the plant process. This means that fault propagation from one CGU to another is achieved only through the control valves or connection points between these CGUs. The simplified P&ID of the reactor CGU (CGU3) can be viewed in figure 3-21.

9.1.1. HDS Model Representation in UML

For better presentation of the P&ID in the object-oriented model the concept of connection point is introduced. The connection point is used to express both merging of two or more lines or splitting of one line into two or more lines. For example, the connection point CP-2-1 represents the merging of the input from H2 and from CGU3.

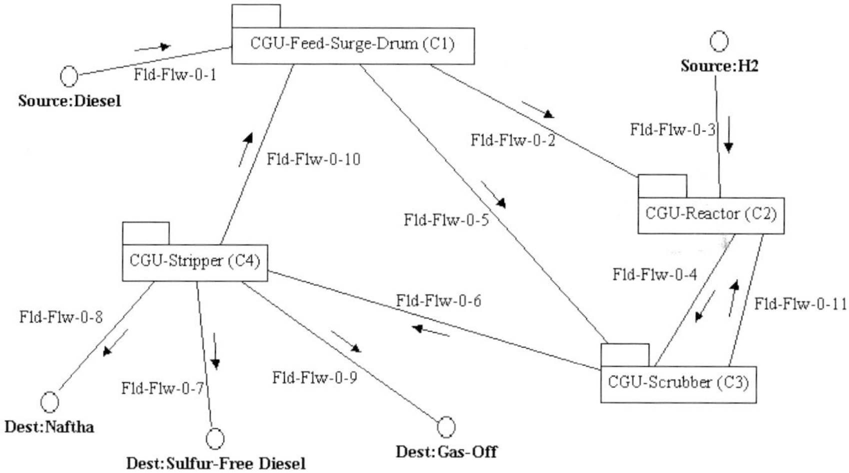

Figure 9-2: HDS Plant CGU-Level Model Representation within UML

HDS plant model is represented using UML in a hierarchical manner. In the top-level of abstraction the plant is considered as a master CGU, which is represented within UML as a package with inputs (Diesel and H2) and outputs (Gas-off, Sulfur-free Diesel, and Naftha). The inputs and outputs of fluid flows are represented as interface classes with sources and destinations connected with the main CGU via fluid flows. Figure 9-2 shows the decomposition of the HDS plant, where CGUs are represented as sub-packages connected by fluid flows.

Fld-Flw-0-1 represents the fluid flow in level (0) from the class "Source:Diesel" to the CGU (C1). Both are referring to the same metaclass with the same properties, although they are in two different levels of abstraction. As a systematic approach, the abstraction level has been used as a part of the naming convention of the fluid flows. Each fluid flow in any level is associated with the following information: (1) "Signal", which is a specification of a asynchronous stimulus between objects and is processed by the StateMachines, (2) "Destination element", which indicates the destination of the fluid flow, and (3) "Metaclass", which links the fluid flow to a given fluid class (defined within the metamodel level).

9.1.2. Reactor CGU Representation within UML

In Figure 9-3, the reactor CGU model is represented using UML where the plant components (i.e. valves, pumps, heat exchange, etc.) are defined as classes within the reactor package. Connections to external packages (CGUs) and inputs/outputs are represented by the connection point (or control valve), which is directly connected to the external CGUs. For example, CGU-Scrubber (C3) has input via CP-2-1, which is merged with H2, and has output via CP-2-6. The behavior of the fluid in the connection points can be done in UML by simply defining the functional and behavioral models and associate them with the connection point interface class, which can not be represented within the P&ID.

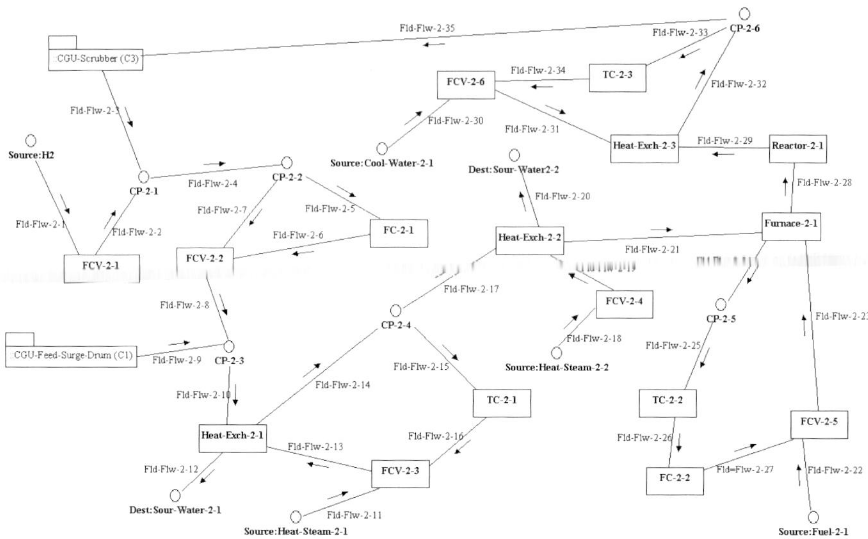

Figure 9-3: Reactor CGU Model Representation in UML

Figure 9-3 shows the detailed model representation of the reactor CGU of HDS plant. In this model, the different components are represented as classes, while the fluid flow is represented as data flow, which is linked to a stereotype of the class of the fluid. For example Fld-Flw-2-28 is the fluid flow from the Furnace-2-1 to the Reactor-2-1. This fluid is associated with the stereotype of the fluid mix of Diesel + H2.

The proposed plant model representation in UML facilitated the conversion of the plant model into HTML format, where plant users can view the plant model and navigate through the different levels of the plant model using normal browser, as shown in figure 9-4.

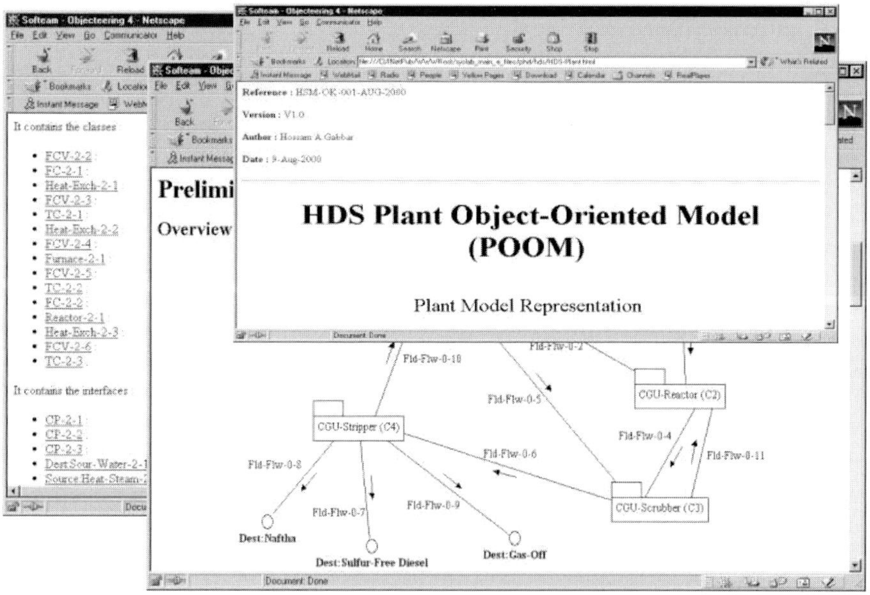

Figure 9-4: HDS Plant Representation in HTML

9.2. Cause – Consequence Analysis of Reactor CGU using CAPE-SAFE

One of the key features of CAPE-SAFE is its ability to automate the hazard evaluation practices using any of the quantitative or qualitative techniques. Also, CAPE-SAFE is used to produce automatically the cause-consequence analysis using ETA technique. Our proposed approach is based on running HAZOP module first to identify the critical top events, which is considered as the initial events of the ETA. The program goes through the underline process, i.e. reactor unit of HDS plant, navigating through the corresponding plant model and identify all the possible fault propagation paths. Scenarios of the fault propagation within the reactor unit are listed in appendix (5).

9.3. Examples from PVC Plant

In this example, PVC batch plant is used as shown in figure 9-5 while table 9-1 shows the operation details of PVC – reaction process.

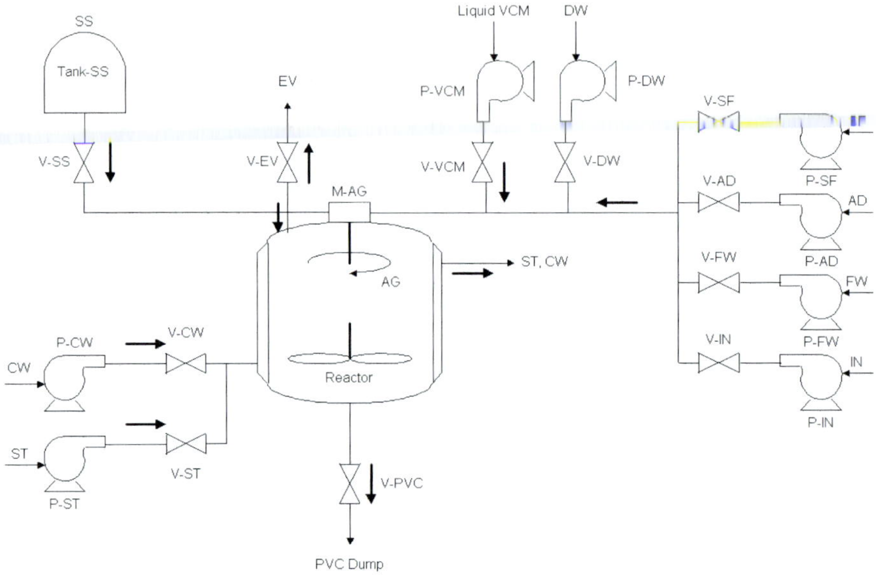

Figure 9-5: Simplified PVC P&ID

Table 9-1: Reactor Operation Steps

1.	After the completion of one batch, reactor vessel is rinsed by charging the reactor with a controlled amount of demineralized water and some surfactants and other additives.
2.	Another batch is started by adding a measured amount of Vinyl Chloride Monomer (VCM).
3.	The polymerization is started by adding a small amount of “initiator” to the batch and steam is circulated through the reactor jacket to heat the contents up to about 327 [K].
4.	The polymerization reaction proceeds exothermically and the steam is replaced by cooling water to maintain the desired reaction temperature of 333 [K].
5.	Agitator inside the vessel keeps the contents well mixed.
6.	After 8 hours, the pressure inside the reactor starts to drop, signaling that most of the VCM has been consumed.

7. The processed batch is then dumped to a downstream facility for further processing

9.3.1. Reactor State Representation

As per ANSI/ISA-S88.01-1995, batch process can be divided into simple operations: preparation, charge, react, and discharge. The corresponding state diagram of the reaction process can be presented in the form of UML, as shown in figure 9-6. In this diagram, each state can be associated with internal transitions. For example, the "preparation state" is associated with an internal transition that receives an event of the completion of previous batch followed by an action of charging the reactor with demineralized water. This is followed by another internal transition of adding the surfactants, which is followed by another internal transition of adding additives. All these internal transitions are within the preparation state. Such state (behavior) model can assist operator while operating the reaction process (Shimada, 2000). The online state can be viewed on the corresponding dynamic web page. In such web pages, the current state and the current internal transition can be highlighted in different color (or blinking). APIs are used to map the operating condition to the plant model and present the results in dynamic web pages.

From figure 9-5, "Reactor is charged [small amount] / Add initiator" is stated as the first line within the "Reaction" state within the UML state diagram. This can be interpreted as follows:

> "Reactor is charged" → is a received event, which triggers the internal transition

"Small amount" → is a guard condition under which an internal transition may be triggered

"Add initiator" → is an action realized when the transition is triggered

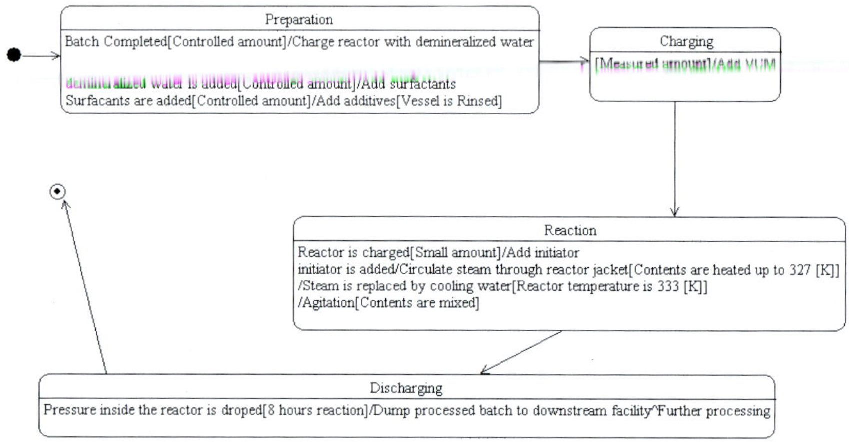

Figure 9-6: Reactor State Diagram Represented in POOM

9.3.2. Representation of Reactor Safety Restrictions in Plant Model

As a part of process safety management programs within batch plants, safety restrictions plays an important role to keep the plant safe during its lifecycle. Within the plant design model, safety restrictions are defined as a constraint associated with the different operations (functions) linked to plant physical components. For example in normal operating condition the control valve (V-IN), which controls the feeding of the initiator, should be opened during the polymerization stage to add controlled amount of the initiator. However, if the reactor temperature exceeds the normal level i.e. abnormal situation, operator should not open such valve to avoid more initiator into the reactor. Such situation can be described as a constraint associated with the valve operation within the plant design model. Operator can view such constraint either in normal situation, or once the temperature exceeds the limits. A popup window within the dynamic web pages is automatically displayed to operator showing the safety restriction associated with the function "open" valve V-IN. Operator can understand why the valve can't (or shouldn't) be open by viewing the details of such constraint within the plant model. The constraints are defined using standard object constraint language (OCL) within the developed plant model. Table 9-2 shows an example constraint associated with the operation OPEN( ) defined for the control valve V-IN.

The constraints are defined in English-like language, which can be easily understood by operator and are graphically represented in the plant model window as popup window associated with the operation OPEN ( ) associated with the control valve V-IN.

Table 9-2: Example of Safety Restrictions Representation Within The Plant Model

```
Object.V-IN.OPEN( ).Constraint {
If Reactor-Temp > React-Temp-Upper
Then
   Open-Permit = False
Else
   Open-Permit = True
}
```

9.3.3. PVC Model Representation in UML

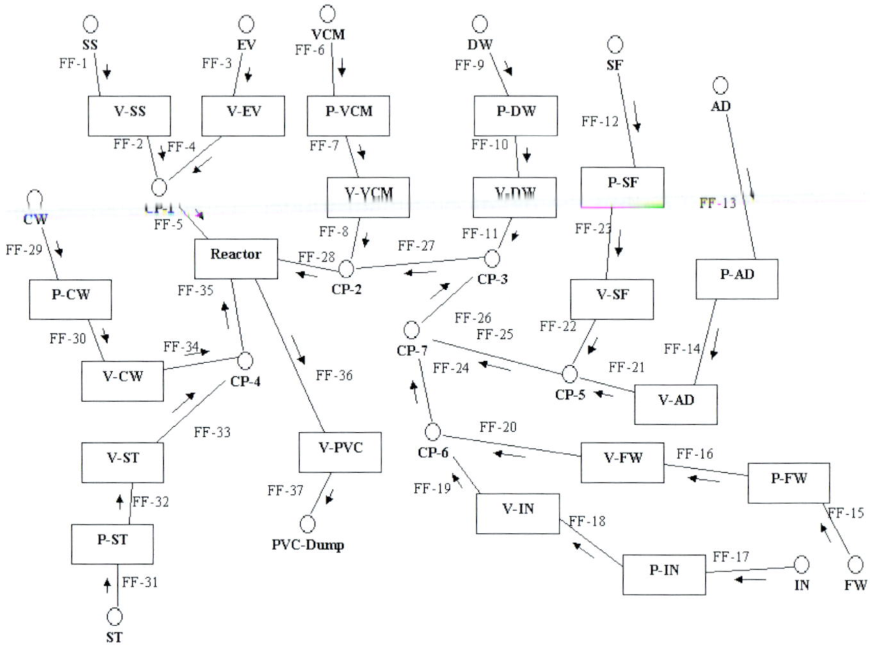

Figure 9-7: PVC Model Representation in UML

PVC model can be represented in UML format as in figure 9-7.

9.4. Examples from Oil Refinery

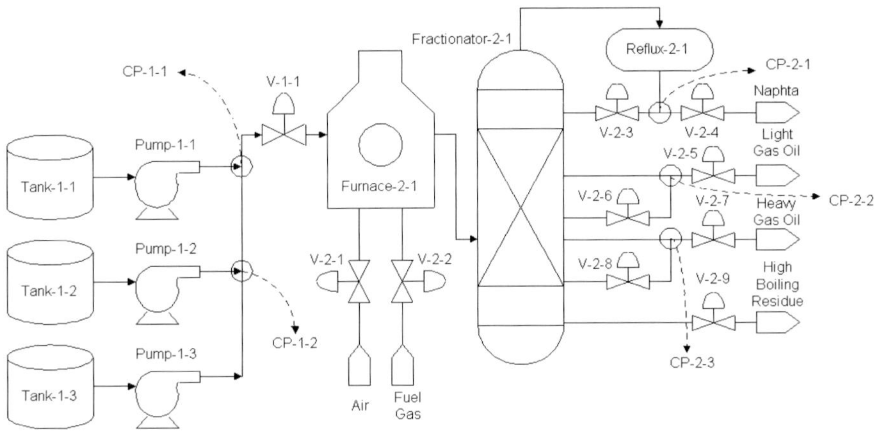

Figure 9-8: Simplified P&ID of Upstream-End of Oil Refinery

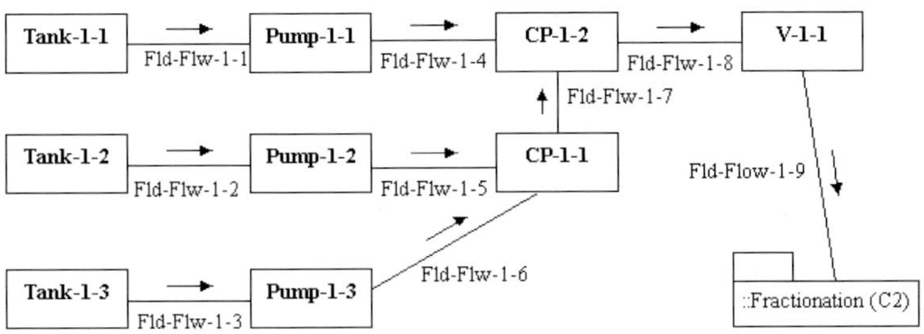

Figure 9-9: Oil Storage Model Representation in UML Format

9.4.1. Oil Storage Process Model in UML

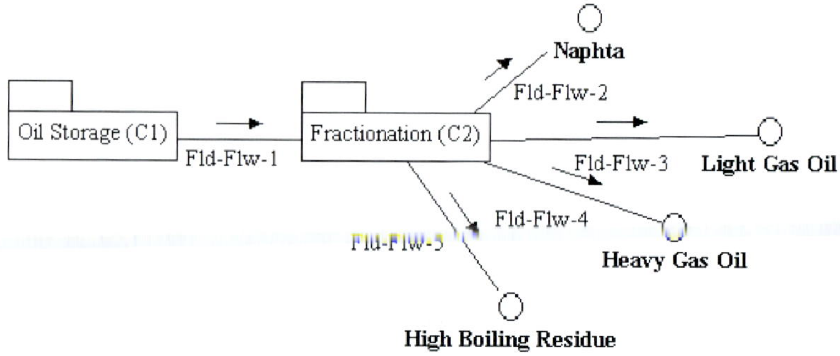

Figure 9-10: Oil Storage CGU Model Representation in UML

Figure 9-8 shows the simplified P&ID of the oil refinery process, while the CGU representation in UML can be shown in figure 9-9. Detailed model of the oil storage CGU (CGU-C1) is shown in figure 9-10. Generic safety procedures are defined as associated with Tank, Pump, connection point, while more specified safety procedures are specified in the level of Pump-1-1, Tank-1-1, etc.

9.4.2. Fractionation Process Model in UML

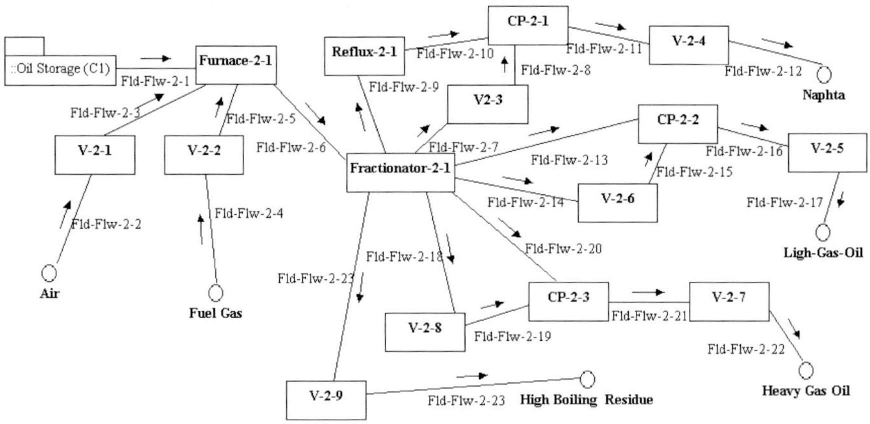

Figure 9-11: Fractionation Process Model Representation in UML Format

Figure 9-11 shows the fractionation process model representation in UML.

9.4.3. Fault Tree Analysis of Oil Refinery

CAPE-SAFE can be used to conduct fault tree analysis, as a call is initiated from the design environment. The fault tree analysis will be implemented automatically using the plant design model. The resulted fault tree can be viewed as in figure 9-12.

In this figure, the selected initial event can be selected from the design environment, or can be extracted by applying automated HAZOP, which highlight list of top events that may cause hazard. In this case, product flow rate (F*) is low has been selected as the initial event. The result of the fault tree analysis as generated by CARA shows total of six cut sets up to order four. The results of the fault tree are fed back to HE Results database as associated with each plant physical object. The relationships among process variable, fault, component, and failure are used in the automation of generating the fault tree results. The minimum cut sets represents the different scenarios for the occurrence of the top event (output product flow rate is low).

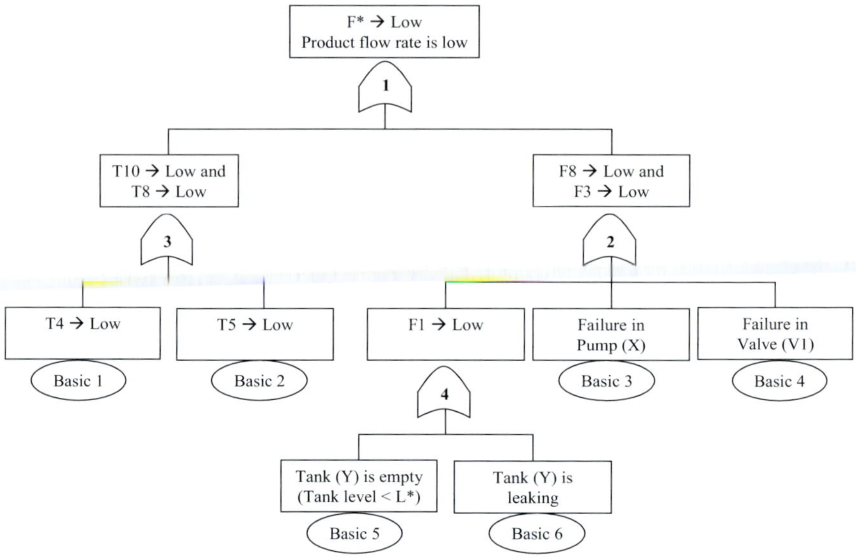

Figure 9-12: Fault Tree Analysis of Oil Refiner Process

9.5. CAPE-SAFE Utilization with Operator Interface System

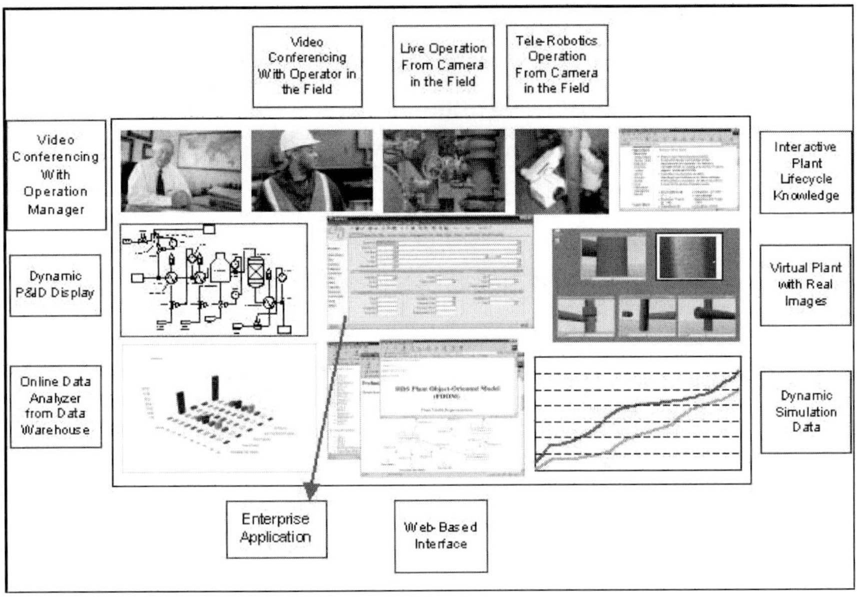

Figure 9-13: Operator Interface System

Figure 9-13 shows a proposed operator interface system (OIS). Such operator interface system helps operator to carry out the daily work by providing information about the plant design and operation. CAPE-SAFE is integrated with the proposed OIS to provide and manipulate safety related data. This is done as per table 9-3.

Table 9-3: CAPE-SAFE interaction with OIS

Interactive Knowledge Window	- Provides information about the reasoning of the failure and provide the root cause of any failure occurring, or simulated by operator, using HEM module
Dynamic P&ID display	- Provides information about the faulty component to be displayed in the
Video conferencing window with operator in the field	- Provides information about the corrective action, or the appropriate next operation to the operator in the

9.6. CAPE-SAFE Utilization with Plant Design Model

Figure 9-14 shows an example of the interaction between HEM within CAPE-SAFE and the design environment CAPE-CAD within the plant enterprise engineering environment. The sequence of activities that will be carried out by HEM are shown where the HDE will decide the suitable hazard evaluation technique based on the selection criteria and process design type. In this figure, the outline structure of the database is proposed. For example, the hazard evaluation results will include the cause, consequence of such hazard, the connected equipment, quantitative results i.e. severity / risk, qualitative results i.e. low / medium / high, and corrective action such as use of alarm / relief valve.

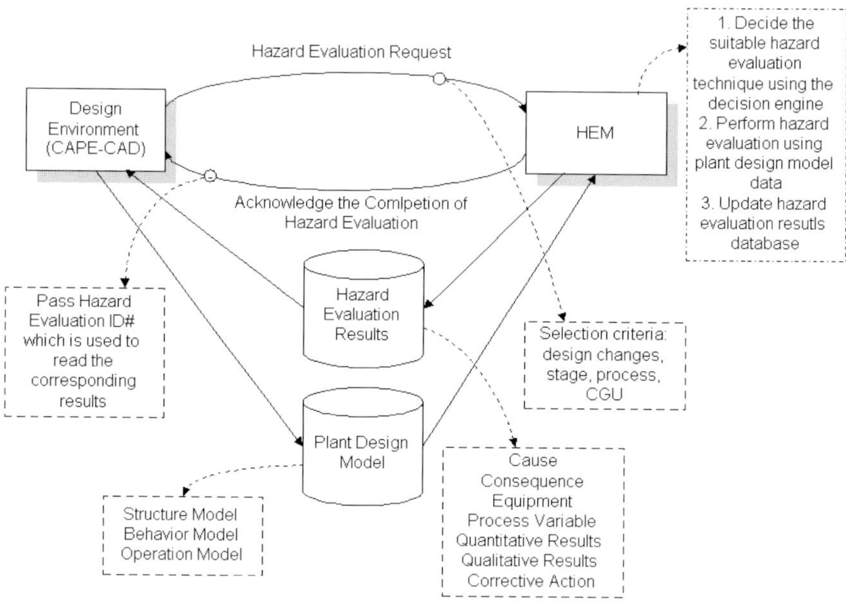

Figure 9-14: Utilization of HEM with Design Environment

9.7. CAPE-SAFE Utilization with Fault Detection System

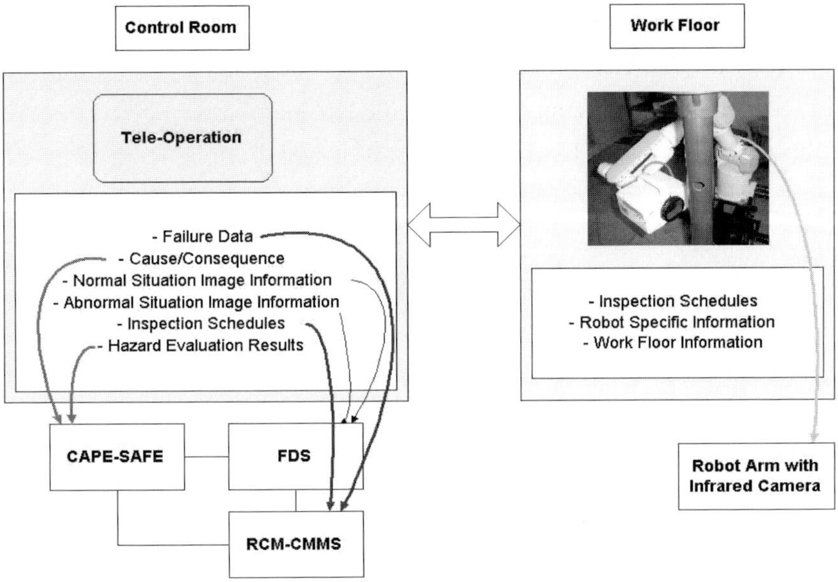

Figure 9-15: CAPE-SAFE Utilization with Fault Detection System

Figure 9-15 shows the utilization of CAPE-SAFE with the fault detection system, which is designed with the use of tele-robotics to detect high temperature or leak in pipes. Such system requires conducting fault analysis for each detected fault to report to operator the root cause and the corrective action. The virtual viewing system display is used, which is also part of the proposed OIS, to display the plant 3D view as well as the actual image of the detected failure. CAPE-SAFE will be acknowledged with the fault and will report the root cause(s) of such failure, which will be acknowledged back to the fault detection system.

9.8. CAPE-SAFE Utilization with RCM-Based CMMS

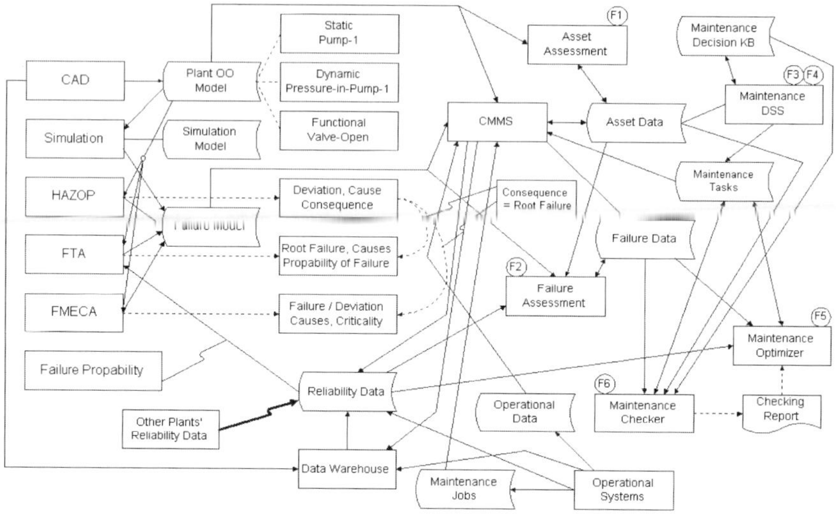

Figure 9-16: RCM-Based CMMS System Architecture

Figure 9-16 shows another example where CAPE-SAFE is integrated with RCM-based CMMS to provide detailed information about the failure model. The safety-related data in the data warehouse; proposed in the RCM-Based CMMS solution; will be maintained from CAPE-SAFE module using predefined database scripts (SQL Server database procedures).

When failure is reported within the maintenance domain, this will initiate a call to CAPE-SAFE to investigate the possible root causes and consequences, which will be reported back to maintenance system. This information is essential to optimize the maintenance work, by identifying the risk and failure details.

Part V: Wrapping Up

10. Discussion

The ultimate goal of achieving safer plant operation has mobilized many researchers to find ways to achieve that. The approach in this book is through modeling the plant lifecycle using the concepts of incremental modeling and metamodeling. Initially, plant design model has been constructed (and can be automated from the CAD environment) in the object-oriented modeling environment in a central repository on the basis of UML. Other plant activities are represented on top of the plant design model (incrementally). For example plant operation model has been represented as associated with the different plant physical objects. The conceptual plant safety model has been designed to identify the ways and opportunities for safety aspects throughout the plant lifecycle to be represented in the plant model. Examples have been illustrated where safety restrictions are represented as safety constraints associated with the different plant model elements suing OCL scripts. Using this approach, safety can be inherited in the plant model, which can ensure safer plant lifecycle i.e. design and operation.

To implement this approach it is required to develop automated modeling environment that maintain the plant model elements and support the representation and manipulation of the different models within the plant enterprise. For that, the book highlighted the design framework of a computer-aided plant enterprise modeling environment called "CAPE-ModE", which is based on the object-oriented modeling and UML.

As safety data, knowledge, and information are highly related to plant design and operation, as well as other activities within the plant enterprise like maintenance, so it is essential to design means of integration among the different components within the plant enterprise.

This has been highlighted as part of the high level specifications of the plant enterprise engineering environment PEEE.

Sides of the implementation of the proposed integrated safety solution have been explained to show how the solution may work in the real plants. This has been shown in the integration with RCM-based CMMS, operation interface system, and the Fault detection system (using tele-robotics technologies).

As stated in the problem statement, the XYZ plant can use CAPE-SAFE as part of its PEEE, where all safety related functions and activities could be performed within CAPE-SAFE.

11. Conclusion

This research work has identified the main components and information system architecture of the plant enterprise engineering environment, which is essential to move to the next generation chemical/petrochemical plants. Within such integrated environment, CAPE-SAFE, as an integrated system to manage all safety aspects within the plant enterprise throughout its lifecycle, has been proposed and designed to manage all safety aspects within plant enterprise throughout its lifecycle. As a part of the specification of CAPE-SAFE, system analysis has been conducted where both contextual and conceptual models have been developed using object-oriented approach. The system architecture and module description of CAPE-SAFE have been specified towards the detailed design and implementation of the proposed system. Plant safety model has been studied and identified to explore all safety aspects and their realization within the plant enterprise engineering environment. Such safety model has been utilized to develop the conceptual design of the proposed CAPE-SAFE.

In order to manage and control the model elements as well as safety representation within the plant model, computer-aided plant enterprise modeling environment (CAPE-ModE) has been suggested. The components and mechanism of CAPE-ModE has been explained. The plant modeling is based on object-oriented modeling approach where safety aspects are defined as associated with each model elements. This achievement can help in automating safety assessment and management practices, especially for complex plants, which will tremendously reduce the operating cost and risk associated with plant operation.

The mechanism of the proposed solution of CAPE-SAFE has been tested on both batch and continuous plant where ways to improve the current practices of safety management have been achieved.

12. Recommendations and Future Research

Currently, plants are struggling to reduce the operating cost while maintaining acceptable level of risk in operating the plant. In addition, national and international health, safety, and environment regulations are forcing plants to follow the standard practices of safety regulations and to conduct regular safety assessment practices. If such requirements are implemented manually they may require longer time and higher cost. The alternative solution is to use automated environment to manage plant lifecycle safety. However, and as stated in this book, it is essential to establish an integrated environment that allow information sharing and data consistency among the different systems and components within the plant enterprise engineering environment. For that it is recommended to implement the complete integrated solution of PEEE using standard and common technologies in the different layers of the information infrastructure. The complete realization of such environment requires prior study to the condition of each plant separately as well as the existing systems. Based on the results of such study, the realization of each component can be recommended.

The development of a detailed conceptual model for the CAPE-SAFE analysis & design stage is a perquisite to start building the complete CAPE-SAFE. The best way of doing that is to use OO CASE as a modeling framework where plant model elements are modeled within the same model pool. The design framework explained in this book is very useful as the base to develop a tailored CAPE-SAFE for other plants. By using unified modeling language it is possible to use or share model elements among different plants, while completing the plant model. This feature is implemented within the CAPE-ModE component to allow import/export and sharing of model elements among different plant models. This includes safety aspects expressed in the plant model.

This means that safety assessment practices can also be enhanced and automated by using last safety assessment results conducted on the same part of the plant and/or utilize safety assessment results conducted in similar plants.

As the idea presented in this book explores an integrated solution for plant enterprise, so more development of new computer-aided systems or components is encouraged that can easily be linked to other components within the environment using the standard proposed technologies i.e. middleware or CORBA.

REFERENCES

Andow, P.K., Lees, F.P.; Trans. Inst. Chem. Engrs 1975, pp 53-195.

Armstrung Laboratory; Information Integration For Concurrent Engineering (IICE) – IDEF4 Object-Oriented Design Method Report Version 2.0. Knowledge Based Systems, Inc. KBSI, Texas, 1995.

Bayer, B., Schneider, R., Marquardt, W., 2000, Integration of data models for process design – first steps and experience. Comp. Chem. Engng., 24, 599-605.

Bieszczad, J., February-2000, A Framework for the language and logic of computer-aided phenomena-based process modeling, Ph.D. Book, MIT.

Bogusch, R. and Marquardt, W., 1997, A formal representation of process model equations, Comp. Chem. Engng., 21, 10, 1105-1115.

Cara™ from Sydvest Software, http://www.sydvest.com/default.htm.

CCPS – Center for chemical process safety. Guidelines for Safe Automation of Chemical Processes. CCPS/AIChE, ISBN 0-8169-0554-1, 1993.

Chikaraishi, M., Suzuki, K.; Development of Dynamic Simulator and Hazard Evaluation System for Safety Design, Master Book in Department of Systems Engineering, Okayama University, Japan, Mar-2000, http://syslab4.mech.okaayama-u.ac.jp/.

Eckel, B., Thinking in Java – Second Edition, http://javafaq.nu/java/book/Contents.shtml, ISBN: 0-13-659723-8.

Elmqvist, H., Erik, S., Otter, M.; Modelica – A Language for Physical System Modeling, Visualization, and Interaction. 1999 IEEE Symposium on Computer-Aided Control System Design, Hawaii, 1999.

G2™ from Gensym, http://www.gensym.com/products/G2.HTM.

Gabbar, H.A. (2001a), Shimada, Y., Suzuki, K.; Operator Interface System Using Plant Object-Oriented Model, Journal of Systems Engineering, Vol. 4, No. 3, 2001.

Gabbar, H.A. (2001b), Suzuki, K., Shimada, Y.; Design Considerations of Computer-Aided RCM-Based Plant Maintenance Management System, ESCAPE-11, Denmark, May-2001, pp 859-864.

Gabbar, H.A. (2001c), Suzuki, K., Shimada, Y.; Manufacturing Process Modeling Using Unified Modeling Language, Conference of Production Engineering, Design, and Control (PEDAC), Alexandria, Egypt, Feb-2001, Vol. I, pp 307-317.

Gabbar, H.A. (2001d), Yamamoto, H., Suzuki, K., Shimada, Y.; Fault Detection System Using Tele-Robotic For Nuclear Power Plant, Conference of Production Engineering, Design, and Control (PEDAC), Alexandria, Egypt, Feb-2001, Vol. II, pp 883-893.

Gabbar, H.A. (2001e), Suzuki, K., Shimada, Y.; Advanced Modeling Methodology For Chemical/Petrochemical Plants, The 7th International Conference On Mining, Petroleum and Metallurgical Engineering (MPM), Assiut, Egypt, Feb-2001, Vol. III-Petroleum, pp 114-120.

Gabbar, H.A. (2000f), Chung, P.W.H., Suzuki, K., Shimada, Y.; Utilization of Unified Modeling Language (UML) to Represent the Artifacts of the Plant Design Model, Conference of Process Systems Engineering (PSE), Kyoto, Japan, Dec-2000, PS54, pp 387-392.

Gabbar, H.A. (2000g), Suzuki, K., Shimada, Y.; Object Oriented Modeling for Process Plant Lifecycle, Proceedings of PSAM5, ISBN 4-946443-64-9, Japan, Nov-2000, Vol. 1, PP: 455-461.

Gabbar, H.A. (2000h), Suzuki, K., Shimada, Y.; Unit Operation Modeling Approach For Plant Safety Practices, Conference Proceedings of Society Of Plant Engineers in Japan (SOPEJ), Nov-2000, No. 3, pp:27–32.

Gabbar, H.A. (2001i), Suzuki, K., Shimada, Y.; Study on Plant Object-Oriented Model Formalization – Case Study: HDS Plant Design,

Journal of Chemical Engineering & Processing, May-2001 (Submitted).

Gabbar, H.A. (2001j); Object-Oriented Modeling of Computer-Aided Plant Enterprise Safety Management System (CAPE-SAFE), 11th International Conference on Computer Theory and Applications ICCTA'2001, Alexandria, Egypt, August-2001.

Gabbar, H.A. (2001k), Suzuki, K., Shimada, Y.; Design Of Process Safety Model In Plant Enterprise Engineering Environment, Reliability Engineering & Safety Systems Journal, Vol. 73, No. 1, pp 35-47, 2001.

Gabbar, H.A. (2001l); Computer-Aided Plant Enterprise Safety Management System (CAPE-SAFE) - Design Framework, Journal of Systems Engineering, Jul-2001 (Submitted).

Gofuku, A. (1994); Representation of Goals-Functions-Structure and Derivation of Behavior for Efficient Design of Engineering System. Institute of Automatic Control System, Technical University of Denmark, ISBN: 87-87950-52.

Hackenberg, J., Krobb, C., Schopfer, G., Wedel, L.v., Wyes, J., and Marquadt, W.; A Repository-Based Environment For Lifecycle Modeling And Simulation. International JSPS Workshop, Tokyo, Japan, 2000.

Halim, I., Palaniappan, C., and Srinivasan, R.; An Integrated Framework for Developing Inherently Safer and Environmentally Benign Processes, Kolding, Denmark, ESCAPE-11, Denmark, May-2001, pp 1145-1150.

ISO, http://www.iso.ch/.

Karvonen, I., Heino, P., Suokas, J.; Knowledge-based approach to support HAZOP studies. Espoo, Techical Research Center of Finland, Research Reports 704, pp 52, 1990.

Kelly, B.E., Lees, F.P. (1986a); The propagation of fault in process plants: 1. Modeling of fault propagation, Journal Reliability Engineering, No. 16, pp 1-38.

Kelly, B.E., Lees, F.P. (1986b); The propagation of fault in process plants: 2. Fault tree synbook, Journal of Reliability Engineering, No. 16, pp 39-62.

Kelly, B.E., Lees, F.P. (1986c); The propagation of fault in process plants: 3. An interactive, computer-based facility, Journal of Reliability Engineering, No. 16, pp 39-62.

Kelly, B.E., Lees, F.P. (1986d); The propagation of fault in process plants: 4. Fault tree synbook of pump system changeover sequence, Journal of Reliability Engineering, No. 16, pp 39-62.

Kimura, F., and Suzuki, H.; Life Cycle Modeling for Innovative Products and Processes, Chapman & Hall, 80-89, 1995.

Linninger, A.A.; Towards Computer-Aided Model Generation, JSPS International Workshop, Tokyo, Japan, 2000.

Linninger, A. A. and H. Krendl, 1999, TechTool – computer-aided generation of process models – part 1 – A generic mathematical language, Comp. Chem. Engng., 23, S703-S706.

Lind, M.; Modeling Goals and Functions of Complex Industrial Plant, Applied Artificial Intelligence, 8:529-283, 1994.

Marquardt, W., 1996, Trends in computer-aided process modeling, Comp. Chem. Engng. 20, 591-609.

Marquardt, W., Bayer, B., Wedel, V. L., 2000, Perspectives on lifecycle process modeling, Foundations of Computer-Aided Process Design, AIChE Symp. Ser. 323, 96, 192-214.

MAXIMO™ from PSDI, www.maximo.com/.

Mizoguchi, R., Gofuku, A., Yoshitsugu; Human Media Interface System for the Next Generation Plant Operation, Tokyo, Japan, IEEE SMC, V-630 – V-635, 1999.

Naftaly H., Minsky, Pal, P.p., Providing Multiple Views for Objects, Software – Practice And Experience, 1-5, 1999.

Naka, Y., Batres, R., and Fuchino, T.; Operational design and its benefits in real-time use, Foundations of computer aided process operations (1999), ISBN 0-8169-0776-5, pp: 570.

Naka Y., Batres R., Adriani A. Operational Design for Startup and Shutdown of Chemical Plants Based on a Topological Approach, AIChE Spring National Meeting, Houston, USA, 1995.

Nelson & Associates. Core Principles of Safety Engineering and Cardinal Rules of Hazard Control, www.hazardcontrol.com/coreprinciples.html, 1996.

NIST – National Institute of Standards and Technology, http://www.nist.gov/.

Object Management Group (OMG), OMG Unified Modeling Language Specification (Version 1.3), **http://www.omg.org/**, 1999.

Occupational Safety & Health Administration (OSHA), http://www.osha.gov/.

Ogunnaike, B.A., Harmon, W.R.; Process Dynamics, Modeling, and Control. Oxford University Press, ISBN 0-19-509119-1, 1994.

PISTEP, http://www.pistep.com/.

Pohjola, V.J.(a); Performance Balance: A New Tool for Environment Conscious Modeling, ENTRÉE'97 Proceedings, France, 1997.

Pohjola, V.J.(b); Performance Balance: A New Tool for Environment Conscious Modeling, ENTRÉE, 1997, France.

Pohjola, V.J.(c); Information Systems Concepts Viewed from Chemical Engineering, IFIP WG 8.1 International Working Conference "ISCO4", Netherlands, 1999.

Pongracz, E., Pohjola, V.; The Conceptual Model of Waste Management, ENTRÉE, 1997, France

Qian, Y., Huang, Q., Lin, W., Li, X.; An object/agent based environment for the computer integrated process operation system, Computers & Chemical Engineering, 24, pp 457-462(2000).

Shilling, J. and Sweeney, P., Three steps to view: Extending the object-oriented paradigm. In OOPSLA '98, 353-361,1989.

Shimada, Y.(a), Gabbar, H.A., Suzuki, K.; Study on Designing the Operation Support System, Conference of PSAM5, Japan, Nov-2000.

Shimada, Y.(b), Gabbar, H.A., Suzuki, K., Naka, Y.; Advanced Decision Method for Plant Operation Support System, Loss Prevention and Safety Promotion in Process Industries, Stockholm, Sweden, June-2001.

Srinivasan, R., Venkatasubramanian, V.; Automating HAZOP analysis of batch chemical plants: Part I. The Knowledge representation framework, Journal of Computers & Chemical Engineering, Vol. 22, No. 9, pp 1345-1355, 1988.

STEP, AP221 Application Handbook, ISO 10303.

STEP Tools Inc. http://www.steptools.com/index.html.

Stephanopoulos, G., Schmitt, W.A.; Systems Biology: an Emergency Theme in Biological Research, Proceedings of European Symposium on Computer Aided Process Engineering – 11, pp 55-68, Denmark, 2001

Umeda, T., Kuriyama, T., O'shima, E., Matsuyama, H. (1980). A graphical approach to cause and effect analysis of chemical processing systems. Chemical Engineering, Vol. 35, pp 2379-2388.

UML Notation guide, OMG, http://cgi.omg.org/uml/.

Van, K.A.L., Rombecher, B.; Safety lifecycle management. Quality and Reliability Engineering International, Vol. 15: 493-500 (1999).

Venkatasubramanian, V., Zhao, J., Viswanathan, S., Zhao, C., and Mu, F.; An Integrated Environment for Batch Process Development – from Recipe to Manufacture, ESCAPE-11, Denmark, May-2001, pp 925-930.

Walker, D. and Tait, R., 2004, Health and safety management in small enterprises: an effective low cost approach, Safety Science, Vol. 42, 69-83.

Wedel, V.L. and W. Marquardt, 2000, ROME: A repository to support the integration of models over the life-cycle of model-based engineering processes, Proc. ESCAPE 10, Florance, (Elsevier) pp. 535-540.

Woods, E.A.; The Hybrid Phenomena Theory, Doctorial Book of Norwegian Institute of Technology, 1993.

Yang, S.H., Stursberg, O., Chung, P.W.H., Kowalewski, S; Automatic Safety Analysis of Computer-Controlled Plants, Computers and Chemical Engineering, 25 (2001), 913-922.

APPENDICES

Appendix (1) – Highlights on UML Standards from OMG

More details about the UML standard specifications can be found in the following web sites:

www.isg.de/people/marc/UmlDocCollection/WhatIsTheUML/whatistheuml.html.
http://omg.com/technology/uml/index.htm.

Appendix (2) – Study on Middleware Technology

Middleware architecture is the replacement of Client/Server architecture. The Middleware is defined as 3-tier (or N-tier), while the traditional Client/Server is defined as 2-tier architecture. In this paper, we will highlight and focus on the Middleware concept rather than the 3-tier architecture.

Definition

Middleware tier is used to manage transactions in real time and mission critical application in distributed environment. The Middleware is the logical base of services of which all applications can use, and provides the ability to separate the business logic from the application structure so that allowing application developer to focus on the business logic not the infrastructure. This separation makes it easy to introduce infrastructure changes across many applications, rather than one at a time. Middleware is a separate process designed to provide the infrastructure service link between clients and servers in a distributed system.

Middleware Software Types

Middleware software comes in different types as per the function needed as follows:

1. Transaction Processing Monitors (TPM)
2. Message-Oriented Middleware (MOM)
3. Object Request Brokers (ORB)

Also, Middleware tier may contain the following types:

- Publish/Subscribe
- Database Replication
- Remote procedure Call
- Database Access Technology
- Object Request Brokers
- Object Monitors
- Transaction Processing Monitors
- Message Queuing
- Message Replication
- Message-Switching Software
- Message Passing
- Transaction Managers

Transaction Processing Middleware (The Pessimist)

Transaction is a logical group of actions through which a change in the state of the business is made (updating data, exchanging funds, etc.). Transaction adhere to the four ACID properties:

- Atomicity
- Consistency
- Isolation
- Durability

Transaction Processing Middleware is designed to protect the integrity of these changes and to provide services for the timely and reliable execution of all processes involved as the result of these changes in a distributed environment. Figure A2-1 shows example of the transaction processing monitor system architecture.

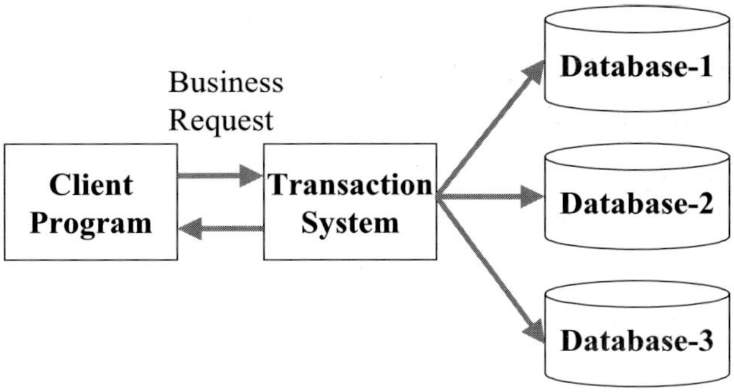

Figure A2-0-1: Transaction Processing Monitor (TPM)

The TPM will protect the transaction from:

- Loss or Corruption
- Systems Failure
- Inaccurate or Incomplete Information
- System Collapses
- Information from Multiple Disparate Systems

Two Phase Commit (2pc)

The 2PC protocol enables a single transaction to update multiple resources managers while guaranteeing integrity.

Message-Oriented Middleware (The Gossips)

Message is a sequence of bytes, which represents a unit of work to the application that send and receives it. Every application has good information to share with others, and crucial to perform a business task. Any type of application-to-application interaction can be achieved by using the MOM. MOMs also provide a separation of the application from the communication protocol allowing interoperability across heterogeneous environments, as shown in figure A2-2.

The following application communication needs to be provided by MOM software:

- Messages That Require a Reply

- Messages That Do Not Require a Reply
- Messages That Need Sharing for Undetermined Reasons
- Messages Broadcast to a Specific Type Application
- Messages Shared Only in the Event Something Happens
- All Messages Must Have Guaranteed Delivery

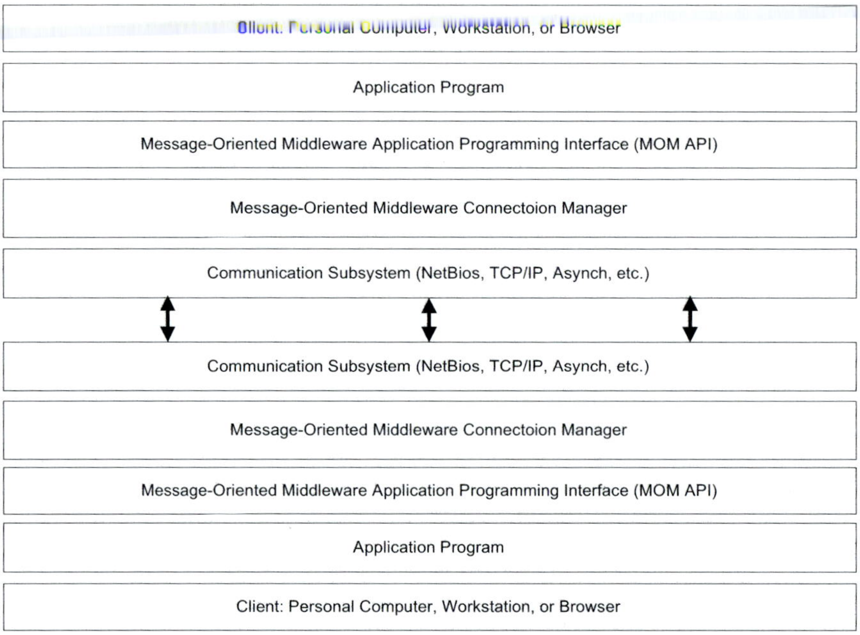

Figure A2-0-2: Base Message-Oriented Middleware Function

Object Request Brokers (Optimist)

Object is a logical self-contained entity containing both data and operations, which can be performed on said data. Objects support the notions of inheritance, encapsulation, and polymorphism. Applications should be designed through the combination of many objects to create a complete application. The ORB is very passive, the object initiating the request is the one in control. Figure A2-3 shows the mechanism of ORB as a flow of requests and responses among objects.

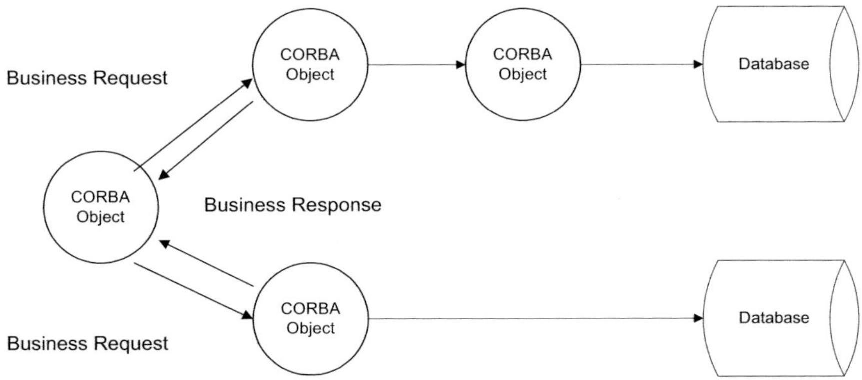

Figure A2-0-3: Object Request Broker

ORB is designed to allow both object and non-object-oriented resources to transparently send and receive requests and responses, independent of language, operating system, and location.

This is possible through an ORB's object standards support and adherence to object-oriented principles.

The main benefits of the ORB are:

- Reuse
- Abstraction

ORB follows the following rules:

1. All objects must be able to communicate with each other
2. Any system, application, protocol, or process should be able to communicate with an object
3. Anything that can be reused, should be
4. All objects have the right to privacy
5. An object invoked by the same method name, need not react the same
6. Each object has the right to choose its own services

Middleware Standards

The standards establish a common method for vendors to deliver Middleware products. There are two major standards:

1. Object Management Group (OMG) standard, which is Common Object Request Broker Architecture (CORBA)
2. Microsoft Standard, which is Distributed Component Object Model (DCOM)

High Level Object Model

Both CORBA and DCOM follow the same high-level object model as follows:

1. Find the object
2. Start the object
3. Connect to the right target object
4. Issue a request
5. Receive a reply

How Objects Communicate?

Not all objects are written in the same language and running on the same machine. So in distributed environment and in order to have interface among all different objects, the only solution is to separate the interface (how the objects communicate with each other) from the implementation (how the object performs its job). This is done using Interface Definition Language (IDL).

Here there is a difference between DCOM and CORBA, Microsoft uses MIDL, while CORBA uses CORBA IDL. Objects talk to each other across the network, using ORB (for CORBA) and Microsoft uses different DCOM products. Multiple CORBA ORBs communicate via a standard communications protocol known as Internet Inter-ORB Protocol (IIOP) as shown in figure A2-4, while Microsoft uses DCOM protocol as shown in figure A2-5. EJB uses its own standard communication for Java-to-Java communication called Remote Method Invocation (RMI). The term BOA is used for "Basic Object Adapter" and RPC is used for "Remote Procedure Call".

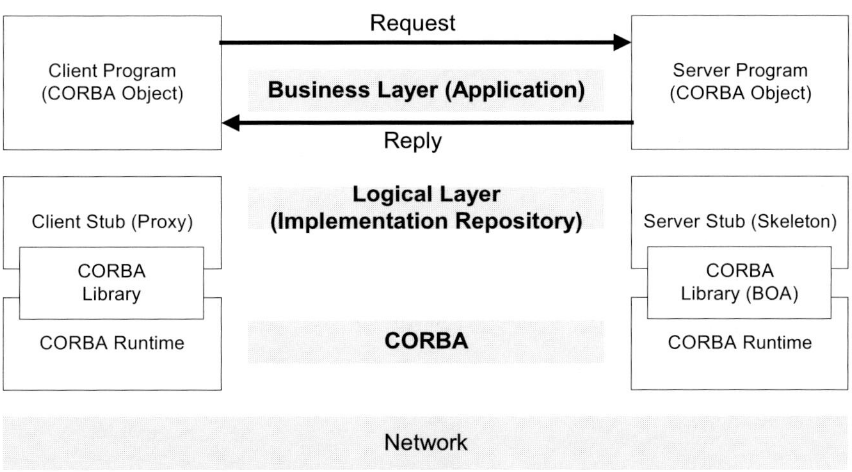

Figure A2-0-4: CORBA communication Model

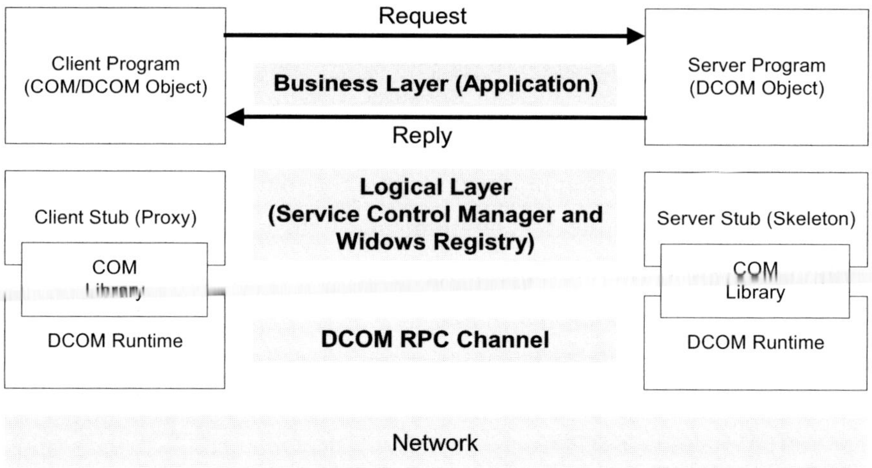

Figure A2-0-5: DCOM communication model

Appendix (3) – Physical Data Model of CAPE-SAFE

In this section the proposed physical data model of CAPE-SAFE will be highlighted as extracted from the database created in SQL Server. In this experiment, MS-Access has been used as a front-end user interface to display and manipulate the data structure created within SQL Server.

Accident Table (Plant_ID, Plant_Location, Accident_ID, Accident_Description, Accident_Causes, Accident_Consequences, Accident_Date, Accident_Participants_ID, Accident_Participant_Role, Linked_Equip_No, Linked_Process_No)

Failure Code Table (Failure_Code, Description)

Failure Hierarchy Table (Failure_List, Failure_Code, Parent)

Failure Remarks In Work Order Table (WO_NO, Description, Entry_Date)

Hazard Table (Hazard_ID, Description, Precaution_Enabled, Hazard_Material_Enabled, Tag_Out_Enabled, Hazard_Type, MSDS-Num, Health-Rating, Flammability_Rating, Reactivity_Rating, Contact_Rating)

Hazard Precaution Table (Hazard_ID, Precaution_ID)

Lock Out Table (Lock_Out_ID, Description, Location, Equip_No, Device_Description, Required_Date)

Precaution Table (Precaution_ID, Description)

Safety Lexicon Table (Safety_Lexicon_ID, Location, Equip_No, Hazard_ID, Tag_Out_ID, Apply_Sequence, Remove_Sequence)

Safety Plan Table (Safety_Plan_ID, Description)

Safety Procedures Table (Safety_Procedure_ID, Description, Linked_Equip_No, Linked_Process_No, Procedure_Condition, Procedure_Action, Linked_Process_Variable, Participant_ID, Linked_Procedure_ID)

Safety Regulation Table (Safety_Regulation_ID, Description, Originator, Linked_Equip_No, Linked_Process_No, Regulation_Condition, Regulation_Action, Linked_Process_Variable, Participant_ID, Linked_Procedure_ID, Assessment_Information, Update_Information)

Safety Training Courses Table (Safety_Course_ID, Course_Title, Description, Course_Objectives, Course_Environment, Course_Duration_In_Days, Course_Prerequisites, Course_Attendees_Category, Exercises_Available)

Safety Training Schedule Table (Safety_Course_ID, Instructor_ID, Month_Day, Hours, Duration)

Safety Training Conducted Courses Table (Safety_Course_ID, Month_Day, Hours, Attendee_ID, Grade)

Tag Lock Table (Tag_Out_ID, Lock_Out_ID, Apply_Sequence, Remove_Sequence, Tag_Lock_ID)

Tag Out Table (Tag_Out_ID, Description, Location, Equip_No, Required_State)

Work Order Hazard Table (WO_NO, Hazard_ID, Description, Precaution_Enabled, Hazard_Material_Enabled, Tag_Out_Enabled, Hazard_Type, MSDS-Num, Health-Rating, Flammability_Rating, Reactivity_Rating, Contact_Rating, WO_Safety_Data_Source)

Work Order Hazard Precaution Table (WO_NO, Hazard_ID, Precaution_ID, WO_Safety_Data_Source)

Work Order Lock Out Table (WO_NO, Lock_Out_ID, Description, Location, Equip_No, Device_Description, Required_State, WO_Safety_Data_Source)

Work Order Precaution Table (WO_NO, Precaution_ID, Description, WO_Safety_Data_Source)

Work Order Precaution Table (WO_Safety_Link_ID, WO_NO, Equip_NO, Location, Hazard_ID, Tag_Out_ID, Apply_Sequence, Remove_Sequence, WO_Safety_Data_Source)

Work Order Safety Plan Table (WO_NO, Safety_Plan_ID, Description)

Work Order Tag Out Table (WO_NO, Tag_Out_ID, Lock_Out_ID, Apply_Sequence, Remove_Sequence, WO_Safety_Data_Source)

Work Order Precaution Table (WO_NO, Tag_Out_ID, Description, Location, Equip_No, Required_Date, WO_Safety_Data_Source)

Appendix (4) – Java Source Code of PEEE

```
import javax.swing.JTabbedPane;
import javax.swing.ImageIcon;
import javax.swing.JLabel;
import javax.swing.JPanel;
import javax.swing.JApplet;
import javax.swing.JFrame;

import javax.swing.*;
import java.awt.*;
import java.awt.event.*;
import javax.swing.JMenu;
import javax.swing.JMenuItem;
import javax.swing.JCheckBoxMenuItem;
import javax.swing.JRadioButtonMenuItem;
import javax.swing.ButtonGroup;
import javax.swing.JMenuBar;
import javax.swing.KeyStroke;
import javax.swing.ImageIcon;
import javax.swing.BorderFactory;
import javax.swing.border.Border;

import javax.swing.JTextArea;
import javax.swing.JScrollPane;
import javax.swing.JFrame;

import javax.swing.JOptionPane;
import javax.swing.JDialog;
import javax.swing.JButton;
```

```
import javax.swing.JToolBar;

import javax.swing.JRadioButton;
import javax.swing.BoxLayout;
import javax.swing.JTabbedPane;
import java.beans.*; //Property change stuff
import javax.swing.JTextField;
import javax.swing.filechooser.*;
import java.io.*;
import java.sql.*;
import java.lang.*;

public class Cape extends JApplet {
    static JFrame frame = new JFrame("CAPEEE");
    static JDialog pwd_w = new JDialog(frame,"CAPEEE:
Login Password");
    static JDialog about_w = new JDialog(frame);
    static JDialog cape_out = new JDialog(frame,"Cape
Output Messages");
    static JDialog cape_in = new JDialog(frame,"Cape Input
Messages");
    static JTextArea display_out,display_in;
    static JScrollPane about_scrollPane;
    static JScrollPane scrollPane_out;
    static JScrollPane scrollPane_in;
    static String simpleDialogDesc = "Some simple message
dialogs";
    static String newline = "\n";
    static Action openAction;
    static Action exitAction;
    static Action aboutAction;
    static JTextField textField;
    static JTextArea textArea;
```

```
    static String selected_file;
    static JFrame fc_frame = new JFrame("File Chooser");
    static JLabel actionLabel;
    static final String useridFieldString = "User-ID";
    static final String passwordFieldString = "Password";
    static String passwordString = "";
    static String useridString;
    static String url = "JDBC:ODBC:PEEE";
    static Connection con;
    static char[] correctPassword = null; //{ 'h', 'o',
's', 's', 'a', 'm' };
    static boolean userFlag = false;
    static boolean useridError = false;
    static boolean userChecked = false;
    static String userid = "";

    public void init() {
    }

    public static void main(String s[]) {

        frame.addWindowListener(new WindowAdapter() {
            public void windowClosing(WindowEvent e) {
                pwd_w.setVisible(false);
                System.exit(0);
            }
        });

        JMenu menu, submenu;
        JMenuItem menuItem;
        JLabel yellowLabel = new JLabel("");
        yellowLabel.setOpaque(true);
        yellowLabel.setBackground(Color.yellow);
```

```
        yellowLabel.setPreferredSize(new Dimension(1020,
747));

        JMenuBar menuBar = new JMenuBar();
        menuBar.setOpaque(true);
        menuBar.setBackground(Color.cyan);
        menuBar.setPreferredSize(new Dimension(1020, 20));

        menu = new JMenu("File");
        menu.setMnemonic(KeyEvent.VK_A);

menu.getAccessibleContext().setAccessibleDescription(
                "The only menu in this program that has
menu items");
        menuBar.add(menu);

        //a group of JMenuItems
        openAction = new AbstractAction("Open File") {
            public void actionPerformed(ActionEvent e) {
                File_Chooser("xx");
            }
        };
        menuItem = new JMenuItem("Open File",
                                        KeyEvent.VK_T);
        menuItem.setAccelerator(KeyStroke.getKeyStroke(
                KeyEvent.VK_1, ActionEvent.ALT_MASK));

menuItem.getAccessibleContext().setAccessibleDescription(
                "Open File");
        menuItem = menu.add(openAction);

        menuItem = new JMenuItem("Close File");
        menuItem.setMnemonic(KeyEvent.VK_B);
```

```
        menu.add(menuItem);

        menuItem = new JMenuItem();
        menuItem.setMnemonic(KeyEvent.VK_D);
        menu.add(menuItem);

        //a submenu
        menu.addSeparator();
        submenu = new JMenu("A submenu");
        submenu.setMnemonic(KeyEvent.VK_S);

        menuItem = new JMenuItem("An item in the
submenu");
        menuItem.setAccelerator(KeyStroke.getKeyStroke(
                KeyEvent.VK_2, ActionEvent.ALT_MASK));
        submenu.add(menuItem);

        menuItem = new JMenuItem("Another item");
        submenu.add(menuItem);
        menu.add(submenu);

        exitAction = new AbstractAction("Exit") {
            public void actionPerformed(ActionEvent e) {
                displayResult("Action for first menu
item", e);
                System.exit(0);
            }
        };
        aboutAction = new AbstractAction("About") {
            public void actionPerformed(ActionEvent e) {
                about_w.setVisible(true);
            }
        };
```

```
        menuItem = new JMenuItem("Exit");
        menuItem.setMnemonic(KeyEvent.VK_B);
        menuItem = menu.add(exitAction);
        menuItem.setIcon(null); //arbitrarily chose not to
use icon in menu

        //Build second menu in the menu bar.
        menuBar.add(Box.createHorizontalGlue());
        menu = new JMenu("Help");
        menu.setMnemonic(KeyEvent.VK_N);

menu.getAccessibleContext().setAccessibleDescription(
                "Help");
        menuBar.add(menu);

        menuItem = new JMenuItem("About");
        menuItem = menu.add(aboutAction);

        frame.setJMenuBar(menuBar);
        frame.getContentPane().add(yellowLabel,
BorderLayout.CENTER);

        ImageIcon icon = new ImageIcon(""); //
images/middle.gif
        JTabbedPane cape_tabbedPane = new JTabbedPane();

        Component capemodel_panel = makeTextPanel("Plant
Enterprise Modeling Environment: CAPE-ModE");
        cape_tabbedPane.addTab("CAPE-ModE", icon,
capemodel_panel, "Plant Enterprise Modeling");
        cape_tabbedPane.setSelectedIndex(0);
```

```
        Component capecad_panel = makeTextPanel("Plant
Design Environment: CAPE-CAD");
        cape_tabbedPane.addTab("CAPE-CAD", icon,
capecad_panel, "Plant Design Environment");

        Component capeoper_panel = makeTextPanel("Plant
Enterprise Operational Systems: CAPE-Oper");
        cape_tabbedPane.addTab("CAPE-Oper", icon,
capeoper_panel, "Plant Operational Systems");



            // CAPE-SAFE tabbed panel
            JTabbedPane capesafe_tabbedPane = new
JTabbedPane();

            Component capesafehem_panel =
makeTextPanel("Hazard Evaluation Manager: HEM");
            capesafe_tabbedPane.addTab("Evaluation (HEM)",
icon, capesafehem_panel, "Hazard Evaluation Manager");

            Component capesafehef_panel =
makeTextPanel("Hazard Evaluation Follow-up: HEF");
            capesafe_tabbedPane.addTab("Follow-up (HEF)",
icon, capesafehef_panel, "Hazard Evaluation Follow-up");

            Component capesafesrm_panel =
makeTextPanel("Safety Regulations Manager: SRM");
            capesafe_tabbedPane.addTab("Regulations
(SRM)", icon, capesafesrm_panel, "Safety Regulations
Manager");

            Component capesafesps_panel =
makeTextPanel("Safety Procedures Manager: SPS");
```

```
            capesafe_tabbedPane.addTab("Procedures (SPS)",
icon, capesafesps_panel, "Safety Procedures Synthesizer");

            JPanel capesafe_panel = new JPanel(true);

            capesafe_panel.setLayout(new GridLayout(1,
1));
            capesafe_panel.add(capesafe_tabbedPane);

            cape_tabbedPane.addTab("CAPE-SAFE", icon,
capesafe_panel, "Plant Enterprise Safety Management");

        Component capesim_panel = makeTextPanel("Plant
Enterprise Simulation Environment: CAPE-Sim");
        cape_tabbedPane.addTab("CAPE-SIM", icon,
capesim_panel, "Plant Simulation");

            // CAPE-ERP tabbed panel
            JTabbedPane capeerp_tabbedPane = new
JTabbedPane();

            Component capeerpcmms_panel =
makeTextPanel("CMMS");
            capeerp_tabbedPane.addTab("CMMS", icon,
capeerpcmms_panel, "Computerized Maintenance Management
System: CMMS");

            Component capeerphr_panel =
makeTextPanel("Human Resources: HR");
            capeerp_tabbedPane.addTab("HR", icon,
capeerphr_panel, "Human Resources: HR");
```

```
            Component capeerpfin_panel =
makeTextPanel("Financial Systems: Fin");
            capeerp_tabbedPane.addTab("FIN", icon,
capeerpfin_panel, "Financial Systems: FIN");

            Component capeerpmnf_panel =
makeTextPanel("Manufacturing: MNF");
            capeerp_tabbedPane.addTab("MNF", icon,
capeerpmnf_panel, "Intelligent Manufacturing System");

            Component capeerpmrkt_panel =
makeTextPanel("Marketting: MRKT");
            capeerp_tabbedPane.addTab("MARKET", icon,
capeerpmrkt_panel, "Product Marketting");

            JPanel capeerp_panel = new JPanel(true);

            capeerp_panel.setLayout(new GridLayout(1, 1));
            capeerp_panel.add(capeerp_tabbedPane);

            cape_tabbedPane.addTab("CAPE-ERP", icon,
capeerp_panel, "Enterprise Resource Planning");

        Component capewf_panel = makeTextPanel("Plant
Enterprise Work Flow Manager: CAPE-WF");
        cape_tabbedPane.addTab("CAPE-WF", icon,
capewf_panel, "Work Flow Management");

            // CAPE-Train tabbed panel
            JTabbedPane capetrain_tabbedPane = new
JTabbedPane();
```

```
            Component capetrainsafe_panel =
makeTextPanel("Safety Training: SAF-TRN");
            capetrain_tabbedPane.addTab("SAF-TRN", icon,
capetrainsafe_panel, "Safety Training: SAF-TRN");

            Component capetrainoper_panel =
makeTextPanel("Plant Operation Training: OPR-TRN");
            capetrain_tabbedPane.addTab("OPR-TRN", icon,
capetrainoper_panel, "Plant Operation Training: OPR-TRN");

            Component capetrainmnt_panel =
makeTextPanel("Plant Maintenance Training: MNT-TRN");
            capetrain_tabbedPane.addTab("MNT-TRN", icon,
capetrainmnt_panel, "Maintenance Training: MNT-TRN");

            Component capetrainadm_panel =
makeTextPanel("Administration Training: ADM-TRN");
            capetrain_tabbedPane.addTab("ADM-TRN", icon,
capetrainadm_panel, "Plant Administration Training: ADM-
TRN");

            Component capetrainfin_panel =
makeTextPanel("Financial Training: FIN-TRN");
            capetrain_tabbedPane.addTab("FIN-TRN", icon,
capetrainfin_panel, "Plant Financial Training: FIN-TRN");

            JPanel capetrain_panel = new JPanel(true);

            capetrain_panel.setLayout(new GridLayout(1,
1));
            capetrain_panel.add(capetrain_tabbedPane);
```

```
            cape_tabbedPane.addTab("CAPE-Train", icon,
capetrain_panel, "Training Management");

        Component capemanager_panel = makeTextPanel("Plant
Enterprise Manager: CAPE-Manager");
        cape_tabbedPane.addTab("CAPE-Manager", icon,
capemanager_panel, "Plant Enterprise Management");

        Component capepsp_panel = makeTextPanel("Plant
Service Provider: CAPE-PSP");
        cape_tabbedPane.addTab("CAPE-PSP", icon,
capepsp_panel, "Plant Service Provider");

        Component capeadmin_panel = makeTextPanel("Plant
Enterprise System Administrator: CAPE-Admin");
        cape_tabbedPane.addTab("CAPE-Admin", icon,
capeadmin_panel, "System Administration");

        //Add the tabbed pane to this panel.
        frame.getContentPane().add(cape_tabbedPane);

        JTextArea about_text = new JTextArea(
              "\n" +
              "\n         (CAPEEE) Computer-Aided Plant
Enterprise Engineering Environment\n" +
              "              is an Integrated Software for
Chemical/Petrochemical\n" +
              "              and Manufacturing
Organizations.\n" +
              "\n" +
              "\n" +
              "              Copyright @ 2001 - System
Analysis Lab., Okayama University\n" +
```

```
            "\n" +
            "                                    Written By:
Hossam A.Gabbar\n");
        about_text.setEditable(false);
        about_scrollPane = new JScrollPane(about_text);
        about_w.getContentPane().add(about_scrollPane,
BorderLayout.NORTH);

        JButton button = null;
        button = new JButton("OK");
        button.setToolTipText("Press OK to return back to
CAPEEE");
        button.addActionListener(new ActionListener() {
            public void actionPerformed(ActionEvent e) {
                about_w.setVisible(false);
            }
        });
        about_w.getContentPane().add(button,
BorderLayout.SOUTH);

        about_w.setSize(450,250);
        about_w.setLocationRelativeTo(frame);
        about_w.setVisible(false);

        JLabel label_pwd = new JLabel("Enter your
password: ");

        final JTextField useridField = new JTextField(10);
        useridField.setActionCommand(useridFieldString);

        //Create a password field.
        final JPasswordField passwordField = new
JPasswordField(6);
```

```
passwordField.setActionCommand(passwordFieldString);
        passwordField.setEchoChar('#');

        //Create some labels for the fields.
        JLabel useridFieldLabel = new
JLabel(useridFieldString + ": ");
        useridFieldLabel.setLabelFor(useridField);
        JLabel passwordFieldLabel = new
JLabel(passwordFieldString + ": ");
        passwordFieldLabel.setLabelFor(passwordField);

        //Create a label to put messages during an action
event.
        actionLabel = new JLabel("Enter your user id and
password.");

actionLabel.setBorder(BorderFactory.createEmptyBorder(10,0
,0,0));

        //Lay out the text controls and the labels.
        JPanel pwdControlsPane = new JPanel();
        GridBagLayout gridbag = new GridBagLayout();
        GridBagConstraints c = new GridBagConstraints();

        pwdControlsPane.setLayout(gridbag);

        JLabel[] labels = {useridFieldLabel,
passwordFieldLabel};
        JTextField[] textFields = {useridField,
passwordField};
        addLabelTextRows(labels, textFields, gridbag,
pwdControlsPane);
```

```
        c.gridwidth = GridBagConstraints.REMAINDER; //last
        c.anchor = GridBagConstraints.WEST;
        c.weightx = 1.0;
        gridbag.setConstraints(actionLabel, c);
        pwdControlsPane.add(actionLabel);
        pwdControlsPane.setBorder(
                BorderFactory.createCompoundBorder(

BorderFactory.createTitledBorder("Login Fields"),

BorderFactory.createEmptyBorder(5,5,5,5)));

        JPanel pwdPane = new JPanel();
        BoxLayout pwdBox = new BoxLayout(pwdPane,
BoxLayout.Y_AXIS);
        pwdPane.setLayout(pwdBox);
        pwdPane.add(pwdControlsPane);

        pwd_w.getContentPane().add(pwdPane);
        pwd_w.setSize(250,160);
        cape_out.setSize(511,100);
        cape_in.setSize(511,100);
        cape_in.setResizable(true);
        cape_out.setResizable(true);
        cape_out.setLocation(0,647);
        cape_in.setLocation(511,647);
        pwd_w.setLocationRelativeTo(frame);

        display_out = new JTextArea(450, 70);
        display_out.setEditable(false);
        scrollPane_out = new JScrollPane(display_out);
```

```
        cape_out.getContentPane().add(scrollPane_out,
BorderLayout.CENTER);
        cape_out.setDefaultCloseOperation(
              JDialog.DO_NOTHING_ON_CLOSE);

        display_in = new JTextArea();
        display_in.setEditable(true);
        scrollPane_in = new JScrollPane(display_in);
        JToolBar toolBar_in = new JToolBar();
        toolBar_in.setFloatable(false);
        JPanel contentPane_in = new JPanel();
        contentPane_in.setLayout(new BorderLayout());
        contentPane_in.setPreferredSize(new Dimension(450,
70));
        contentPane_in.add(toolBar_in,
BorderLayout.NORTH);
        contentPane_in.add(scrollPane_in,
BorderLayout.CENTER);
        addButtons(toolBar_in);
        cape_in.getContentPane().add(contentPane_in);
        cape_in.setDefaultCloseOperation(
              JDialog.DO_NOTHING_ON_CLOSE);

        pwd_w.addWindowListener(new WindowAdapter() {
            public void windowClosing(WindowEvent e) {
                frame.setVisible(false);
                System.exit(0);
            }
        });
//                  cape_out.setVisible(true);
//                  cape_in.setVisible(true);
//                  frame.setVisible(true);
```

```
    useridField.addFocusListener(new FocusListener() {
        public void focusLost(FocusEvent e) {
          if (!useridError) {
              userid = useridField.getText();
              String userid = useridField.getText();
              validateUserID();
              if (!userFlag) {
                    useridError = true;


  JOptionPane.showMessageDialog(pwd_w,
                                    "Invalid user id
focus. Try again.",
                                    "Error Message",

  JOptionPane.ERROR_MESSAGE);
                   userFlag = true;
                   useridField.setText("");
                   passwordField.setText("");
                     useridError = false;
                   useridField.requestFocus();
              }
          } else {
              useridError = false;
              useridField.requestFocus();
          }
        }
        public void focusGained(FocusEvent e) {
              useridError = false;
        }
    });
```

```
        useridField.addActionListener(new ActionListener()
{
            public void actionPerformed(ActionEvent e) {
            System.out.println("\n action of user\n");
                  useridError = true;
                  userid = useridField.getText();
                  validateUserID();
                  if (!userFlag) {

JOptionPane.showMessageDialog(pwd_w,
                                    "Invalid user id. Try
again.",
                                    "Error Message",

      JOptionPane.ERROR_MESSAGE);
                        userFlag = true;
                        useridField.setText("");
                        passwordField.setText("");
                        useridField.requestFocus();
                        useridError = false;
                  }
                  else {
                        useridError = false;
            System.out.println("go to password field\n");
                        passwordField.requestFocus();
                  }
            }
        });

        passwordField.addFocusListener(new FocusListener()
{
            public void focusGained(FocusEvent e) {
                  if (!userFlag) {
```

```
                    useridError = false;
                    useridField.requestFocus();
                }
            }
            public void focusLost(FocusEvent e) {
                useridError = false;
            }
        });

        passwordField.addActionListener(new
ActionListener() {
            public void actionPerformed(ActionEvent e) {
                useridError = false;
                JPasswordField input =
(JPasswordField)e.getSource();
                char[] password = input.getPassword();
                if (isPasswordCorrect(password)) {
                    pwd_w.setVisible(false);
                    cape_out.setVisible(true);
                    cape_in.setVisible(true);
                    frame.setVisible(true);
                } else {
                    JOptionPane.showMessageDialog(pwd_w,
                        "Invalid password. Try again.",
                        "Error Message",
                        JOptionPane.ERROR_MESSAGE);
                        passwordField.setText("");
                        passwordField.requestFocus();
                }
            }
        });

        pwd_w.setVisible(true);
```

```
        frame.pack();
        frame.setSize(1022, 647);
    }

    static void File_Chooser(String xx) {
        // select file using filechooser
        final JTextArea log = new JTextArea(5,20);
        log.setMargin(new Insets(5,5,5,5));
        log.setEditable(false);
        JScrollPane logScrollPane = new JScrollPane(log);

        //Create a file chooser
        final JFileChooser fc = new JFileChooser();

        //Create the open button
        ImageIcon openIcon = new
ImageIcon("images/open.gif");
        JButton openButton = new JButton("Open a File...",
openIcon);
        openButton.addActionListener(new ActionListener()
{
            public void actionPerformed(ActionEvent e) {
                int returnVal =
fc.showOpenDialog(fc_frame);

                if (returnVal ==
JFileChooser.APPROVE_OPTION) {
                    File file = fc.getSelectedFile();
                    //this is where a real application
would open the file.
                    log.append("Opening: " +
file.getName() + "." + newline);
                    selected_file = file.getName();
```

```
                } else {
                    log.append("Open command cancelled by
user." + newline);
                    selected_file = "";
                }
            }
        });

        //Create the save button
        ImageIcon saveIcon = new
ImageIcon("images/save.gif");
        JButton saveButton = new JButton("Save a File...",
saveIcon);
        saveButton.addActionListener(new ActionListener()
{
            public void actionPerformed(ActionEvent e) {
                int returnVal =
fc.showSaveDialog(fc_frame);
                if (returnVal ==
JFileChooser.APPROVE_OPTION) {
                    File file = fc.getSelectedFile();
                    //this is where a real application
would save the file.
                    log.append("Saving: " + file.getName()
+ "." + newline);
                    selected_file = file.getName();
                } else {
                    log.append("Save command cancelled by
user." + newline);
                    selected_file = "";
                }
            }
        });
```

```
        //Create the close button
        JButton closeButton = new JButton("Close File
Chooser");
        closeButton.addActionListener(new ActionListener()
{
            public void actionPerformed(ActionEvent e) {
                  selected_file = log.getText();
                  log.setText("");
                  display_out.append(selected_file);
                  fc_frame.setVisible(false);
            }
        });

        //For layout purposes, put the buttons in a
separate panel
        JPanel buttonPanel = new JPanel();
        buttonPanel.add(openButton);
        buttonPanel.add(saveButton);
        buttonPanel.add(closeButton);

        //Explicitly set the focus sequence.
        openButton.setNextFocusableComponent(saveButton);
        saveButton.setNextFocusableComponent(openButton);

        //Add the buttons and the log to the frame
        JPanel fccontentPane = new JPanel();
        fccontentPane.add(buttonPanel,
BorderLayout.NORTH);
        fccontentPane.add(logScrollPane,
BorderLayout.CENTER);
        fc_frame.setSize(800,300);
        fc_frame.setLocation(110,100);
```

```
        fc_frame.getContentPane().add(fccontentPane);

        fc_frame.addWindowListener(new WindowAdapter() {
            public void windowClosing(WindowEvent e) {
                selected_file = log.getText();
                log.setText("");
                display_out.append(selected_file);
                fc_frame.setVisible(false);
            }
        });
        fc_frame.setVisible(true);
        return;
    }

    static void addLabelTextRows(JLabel[] labels,
                                 JTextField[] textFields,
                                 GridBagLayout gridbag,
                                 Container container) {
        GridBagConstraints c = new GridBagConstraints();
        c.anchor = GridBagConstraints.EAST;
        int numLabels = labels.length;

        for (int i = 0; i < numLabels; i++) {
            c.gridwidth = GridBagConstraints.RELATIVE;
//next-to-last
            c.fill = GridBagConstraints.NONE;      //reset
to default
            c.weightx = 0.0;                       //reset
to default
            gridbag.setConstraints(labels[i], c);
            container.add(labels[i]);
            c.gridwidth = GridBagConstraints.REMAINDER;
//end row
```

```
            c.fill = GridBagConstraints.HORIZONTAL;
            c.weightx = 1.0;
            gridbag.setConstraints(textFields[i], c);
            container.add(textFields[i]);
        }
    }

    static void addButtons(JToolBar toolBar) {
        JButton button = null;

        //execute button
        button = new JButton(new
ImageIcon("images/left.gif"));
        button.setToolTipText("Execute user command in the
input window");
        button.addActionListener(new ActionListener() {
            public void actionPerformed(ActionEvent e) {
                executeUserCommand(display_in.getText(),
e);
            }
        });
        toolBar.add(button);

        //clear button
        button = new JButton(new
ImageIcon("images/right.gif"));
        button.setToolTipText("Clear input and output
windows");
        button.addActionListener(new ActionListener() {
            public void actionPerformed(ActionEvent e) {
                display_in.setText("");
                display_out.setText("");
            }
```

```
        });
        toolBar.add(button);
    }

    protected static Component makeTextPanel(String text)
{
        JPanel panel = new JPanel(false);
        JLabel filler = new JLabel(text);
        filler.setHorizontalAlignment(JLabel.CENTER);
        panel.setLayout(new GridLayout(1, 1));
        panel.add(filler);
        return panel;
    }

    private static boolean isPasswordCorrect(char[] input)
{
//    display_out.append("\npassword is space " +
correctPassword.toString() + " input password " +
input.toString());
        if (input.length != correctPassword.length) {
            return false;
        }
        for (int i = 0;  i < correctPassword.length; i ++)
{
            if (input[i] != correctPassword[i]) {
                return false;
            }
        }
        return true;
    }

    static void displayResult(String actionDescription,
                                    ActionEvent e) {
```

```
        String s = ("Action event detected by: "
                   + actionDescription
                   + newline
                   + "    Event source: " + e.getSource()
                   + newline);
        display_out.append(s);
    }

    static void executeUserCommand(String userCommand,
                                 ActionEvent e) {
        String s = ("User Command:> " + userCommand +
newline);
        display_out.append(s);
    }

    private static boolean checkForWarning (SQLWarning
warn)

      throws SQLException  {
      boolean rc = false;

      // If a SQLWarning object was given, display the
      // warning messages.  Note that there could be
      // multiple warnings chained together

      if (warn != null) {
            System.out.println ("\n *** Warning ***\n");
            rc = true;
            while (warn != null) {
                  System.out.println ("SQLState: " +
                        warn.getSQLState ());
                  System.out.println ("Message:  " +
                        warn.getMessage ());
```

```
                    System.out.println ("Vendor:   " +
                        warn.getErrorCode ());
                    System.out.println ("");
                    warn = warn.getNextWarning ();
                }
            }
            return rs;
        }

        private static void dispResultSet (ResultSet rs)
            throws SQLException {
            int i;

            // Get the ResultSetMetaData.  This will be used for
            // the column headings

            ResultSetMetaData rsmd = rs.getMetaData ();

            // Get the number of columns in the result set

            int numCols = rsmd.getColumnCount ();

            // Display data, fetching until end of the result
set
            userFlag = rs.next();
            if (userFlag) {
                passwordString = rs.getString(1);
                passwordString = passwordString.trim();
                correctPassword =
passwordString.toCharArray();
            }
        }
```

```
   private static void validateUserID () {
                 correctPassword = null;
                 passwordString = null;
                 //ResultSet rs = null;
                 try {
                       // Load the jdbc-odbc bridge
driver

      Class.forName("com.ms.jdbc.odbc.JdbcOdbcDriver");
                 }
                 catch(java.lang.ClassNotFoundException
ex) {

      display_out.append("ClassNotFoundException: ");

display_out.append(ex.getMessage());
                 }
                 try {
                       // Attempt to connect to a driver.
                       con = DriverManager.getConnection
(
                             url, "sa", "hsmsyslab");
                       checkForWarning (con.getWarnings
());

                       String query = "SELECT password
FROM plantusers"
                             + " WHERE userid = " +
userid;

                       Statement stmt =
con.createStatement();
```

```
                        // Submit a query, creating a
ResultSet object
                        ResultSet rs = stmt.executeQuery
(query);

                        dispResultSet (rs);

                        // Close the statement
                        stmt.close();
                        // Close the connection
                        con.close();

                        // Close the result set
                        rs.close();
                  }
                  catch (SQLException ex) {
                        // A SQLException was generated.
Catch it and
                        // display the error information.
Note that there
                        // could be multiple error objects
chained
                        // together

                        display_out.append("\n***
SQLException caught ***\n");

                        while (ex != null) {

      display_out.append("SQLState: " +
                              ex.getSQLState ());
                              display_out.append("Message:
" + ex.getMessage ());
```

```
                              display_out.append("Vendor:
" +
                                   ex.getErrorCode ());
                              ex = ex.getNextException ();
                              display_out.append("\n");
                         }
                    }
                    catch (java.lang.Exception ex) {
                         // Got some other type of
exception.  Dump it.
                         ex.printStackTrace();
                    }
          }
}
```

Appendix (5) – Cause/Consequence Scenarios of Reactor Unit in HDS Plant

Table A5-1: Scenario S1 – Cause / Consequence Analysis of Reactor Unit

Cause	Consequence
(1) Failure of TC-2-2, Temperature low	(2) Feed amount of fuel Gas to Furnace-2-1 becomes bigger
(2) Feed amount of fuel Gas to Furnace-2-1 becomes bigger	(3) Abnormal combustion inside Furnace-2-1
	(4) Temperature increase of the liquid mixture Fld-Flw-2-28 (Diesel + H2)
(3) Abnormal combustion inside Furnace-2-1	(5) Breakage of Furnace-2-1
(5) Breakage of Furnace-2-1	(6) Thermal emission to outside (disturbance)
(4) Temperature increase of the liquid mixture Fld-Flw-2-28 (Diesel + H2)	(7) Temperature increase in Reactor-2-1
(7) Temperature increase in Reactor-2-1	(8) Temperature runaway inside Reactor-2-1
	(9) Temperature increase of catalyst (H2) in Reactor-2-1
(8) Temperature runaway inside Reactor-2-1	(10) Fracture of Reactor-2-1
(9) Temperature increase of catalyst (H2) in Reactor-2-1	(11) Possibility of coking in Reactor-2-1
(11) Possibility of coking in Reactor-2-1	(12) Blockage around nozzle
	(13) Loss of pressure in the Reactor-2-1 by coking
(12) Blockage around nozzle	(14) Flow rate decrease of H2 for quench
(14) Flow rate decrease of H2 for quench	(15) Temperature increase in Reactor-2-1 (loop back to 7)

(13) Loss of pressure in the Reactor-2-1 due to coking	(16) Flow rate decrease of H2 due to quench
	(17) Temperature increase of catalyst H2 in Reactor-2-1 (loop back to 9)

Table A5-2: Scenario S2 – Cause / Consequence Analysis of Reactor Unit

Cause	Consequence
(1) Initial event: No flow of steam due to abnormality of steam stripping in Stripper CGU (CGU4)	(2) Overheat of steam heat coil in Furnace-2-1, or breakage of heating tube
	(3) Temperature increase of catalyst H2 in Reactor-2-1
(2) Overheat of steam heat coil in Furnace-2-1, or breakage of heating tube	(4) Fire in Furnace-2-1 or breakage of burner
(4) Fire in Furnace-2-1 or breakage of burner	(5) External heat emission
(3) Temperature increase of catalyst H2 in Reactor-2-1	(6) Coking in Reactor-2-1
(6) Coking in Reactor-2-1	(7) Blockage around nozzle due to coking
	(8) Pressure-loss increase due to coking
(7) Blockage around nozzle due to coking	(9) Decrease of hydrogen for quench
(9) Decrease of hydrogen for quench	(10) Temperature increase in Reactor-2-1
(8) Pressure-loss increase due to coking	(11) Decrease of hydrogen for quench
(11) Decrease of hydrogen for quench	(12) Temperature increase of catalyst H2 in Reactor-2-1

Table A5-3: Scenario S3 – Cause / Consequence Analysis of Reactor Unit

Cause	Consequence
(1) Leak of diesel	(2) Decrease in ratio of H2 in liquid mixture (diesel+H2) in Reactor-2-1
	(3) Leak of diesel to outside reactor CGU (CGU2)
(2) Decrease in ratio of H2 in liquid mixture (diesel+H2) in Reactor-2-1	(4) Drop in temperature in Reactor-2-1
(4) Drop in temperature in Reactor-2-1	(5) Reduce in the inertia of reactor in Reactor-2-1
(5) Reduce in the inertia of reactor in Reactor-2-1	(6) Affect the amine Scrubber CGU (CGU3)

Appendix (6) – Useful Web Links

University of British Columbia - The Department of Health, Safety and Environment	http://www.safety.ubc.ca
Center for Process Systems Engineering	http://www.ps.ic.ac.uk/
Step Tools Ins., "Building Step Solutions for interoperability and e-Manufacturing Web page"	http://www.steptools.com/index.html
University of California Santa Barbara – Chemical Engineering	http://www.chemengr.ucsb.edu/
Mary Kay O'Connor Process Safety Center	http://mkopsc.tamu.edu/
IMS	http://www.ims.org/index2.htm
POEM	http://www.vtt.fi/aut/rm/projects/poem/index .htm
OMG	http://www.omg.com/
UML	http://www.omg.org/uml/
Continuous System Modeling	http://www.ece.arizona.edu/~cellier/springer .html
Nuclear Regulatory Commission	http://www.nrc.gov/OPA/reports/nrc.html
OREDA – Offshore Reliability Data	http://www.sintef.no/sipaa/prosjekt/oreda/in dex.html
BEA – Middleware Solutions	http://www.beasys.com/
Javasoft Java Reference Web page	http://java.sun.com/products//jdk/1.2/docs/a pi/index.html
JavaTM 2 SDK Tools	http://java.sun.com/j2se/1.3/docs/tooldocs/to

and Utilities	ols.html
Okayama University, Department of Systems Engineering, System Analysis Lab.	http://syslab4.mech.okayama-u.ac.jp/
Professor Paul Chung, Loughborough University	http://www.lboro.ac.uk/departments/co/personal_pages/chung.html
The Molecular Modeling Research Group Server At DISAABA	http://antas.agraria.uniss.it/
Georgetown Molecular Modeling Center	http://molmod2.chem.georgetown.edu/
Molecular Modeling for Chemical Education	http://www.molecules.org/

Appendix (7) – Molecular Modeling Impact on CAPE-SAFE

Introduction About Molecular Modeling

Molecular modeling, also known as molecular mechanics, is a method to calculate the structure and energy of molecules based on nuclear motions. Electrons are not considered explicitly, but rather it is assumed that they will find their optimum distribution once the positions of the nuclei are known. This assumption is based on the Born-Oppenheimer approximation of the Schrödinger equation. The Born-Oppenheimer approximation states that nuclei are much heavier and move much more slowly than electrons. Thus, nuclear motions, vibrations and rotations can be studied separately from electrons; the electrons are assumed to move fast enough to adjust to any movement of the nuclei.

In a very crude sense molecular modeling treats a molecule as a collection of weights connected with springs where the weights represent the nuclei and the springs represent the bonds.

Figure A7-0-1: Molecular Model

A force field is used to calculate the energy and geometry of a molecule. It is a collection of atom types (to define the atoms in a molecule), parameters (for bond lengths, bond angles, etc.) and equations (to calculate the energy of a molecule). In a force field a given element may have several atom types. For example, ethylbenzene contains both sp3-

hybridized carbons and aromatic carbons. sp3-Hybridized carbons have a tetrahedral bonding geomtery, while aromatic carbons have a trigonal bonding geometry. The C-C bond in the ethyl group differs from a C-C bond in the phenyl ring, and the C-C bond between the phenyl ring and the ethyl group differs from all other C-C bonds in ethylbenzene. The force field contains parameters for these different types of bonds. Some of these parameters are given below. The total energy of a molecule is divided into several parts called force potentials, or potential energy equations. Force potentials are calculated independently, and summed to give the total energy of the molecule. Examples of force potentials are the equations for the energies associated with bond stretching, bond bending, torsional strain and van der Waals interactions. These equations define the potential energy surface of a molecule.

$$E_{TOTAL} = E_{STRETCH} + E_{BEND} + E_{S\text{-}B} + E_{TORSION} + E_{vdW} + E_{DP\text{-}DP}$$

Below is an example of one of the force potentials, and parameters one may find in a force field. This example is from Allinger's MM2 force field [(a)"Conformational Analysis. 130. MM2. A Hydrocarbon Force Field Utilizing V1 and V2 Torsional Terms", Allinger, N. L., J. Am. Chem. Soc. 1977, 99, 8127. (b) Burket, U.; Allinger, N. L. Molecular Mechanics; American Chemical Society: Washington, DC, 1982.] and the MMX force field of PCMODEL ["PCMODEL", Gilbert, K., Serena Software: Bloomington, IN, 1993].

Energy due to Bond Stretching

Whenever a bond is compressed or stretched the energy goes up. The energy potential for bond stretching and compressing is described by an equation similar to Hooke's law for a spring, except a cubic term is added. This cubic term helps to keep

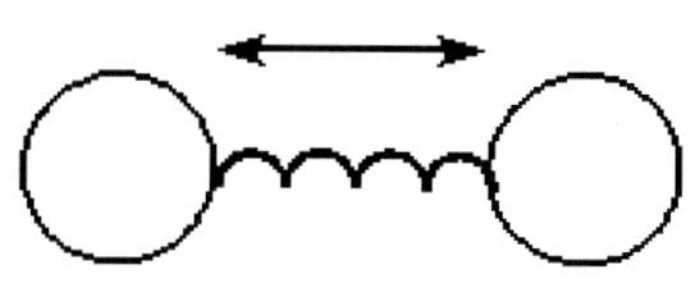

Figure A7-0-2: Energy Due to Bond Stretching

the energy from rising too sharply as the bond is stretched.

Molecular Modeling Application in Chemical Process Safety

By abstracting the chemical/petrochemical processes, one can find that the fluids are a key element where it has interaction with the physical equipment, environment, and humans. The comprehensive study of the fluid behavior is essential to achieve complete solution to any problem related to chemical processes, i.e. process safety. In the reaction process that occur in the reactor within an HDS plant, the fluid inside the reactor goes in different stages and states. One of these states could be abnormal or hazardous condition, where the temperature goes above the limits. To study this phenomena one approach is to use the dynamic simulation, where the behavioral equations that govern the fluid dynamics can be developed and used. Another approach, which goes one step further is to study and develop the molecular models representing the fluid and develop and study the effects on the molecules using simulation. VRML, which is virtual reality molecular language, has been proposed by the researchers to develop 3D models that represent the molecules that can be used also for the molecular simulation practices. Figure A7-3 shows one example representing molecular models.

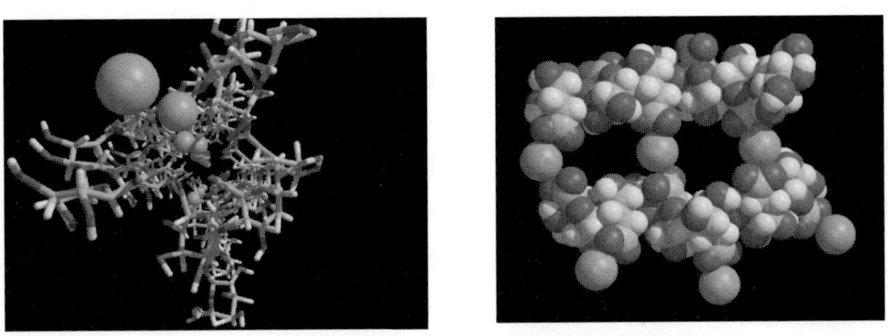

Figure A7-0-3: 3D Examples of Molecular Models

Appendix (8) – Manufacturing Process Modeling

Similar to chemical/petrochemical plants, the manufacturing system architecture will include manufacturing model as integrated with the manufacturing process design environment i.e. CAD/CAM, as shown in figure A8-1.

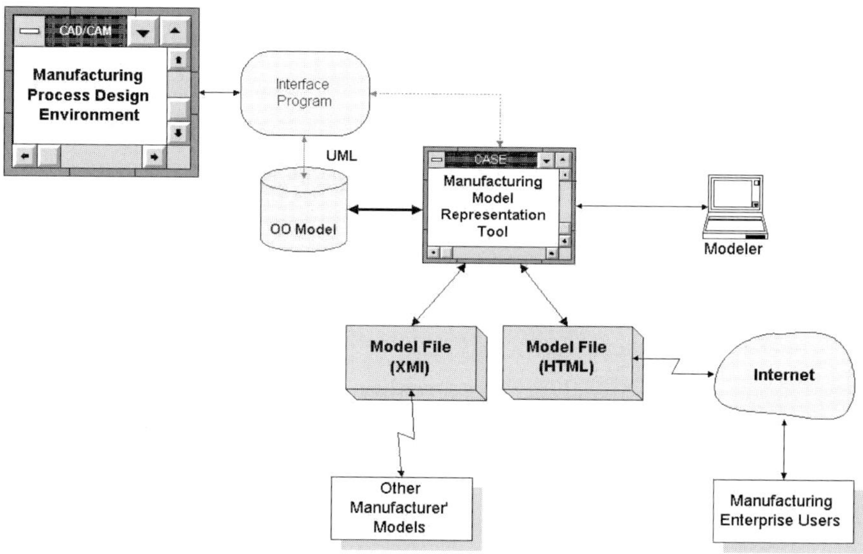

Figure A8-0-1: Manufacturing Modeling System Architecture

The concept of safety can be implemented similarly as realized in the object-oriented model. The control valve manufacturing process, as shown in figures A8-2 and A8-3, will be represented within the manufacturing object-oriented modeling environment on the basis of UML as both the logical points for safety assessment practices i.e. sign directed graph representation, as well as the physical components and sub-components of the control valve, as shown in figure A8-4. By applying the safety aspects in the OO model, safer product and manufacturing process can be achieved.

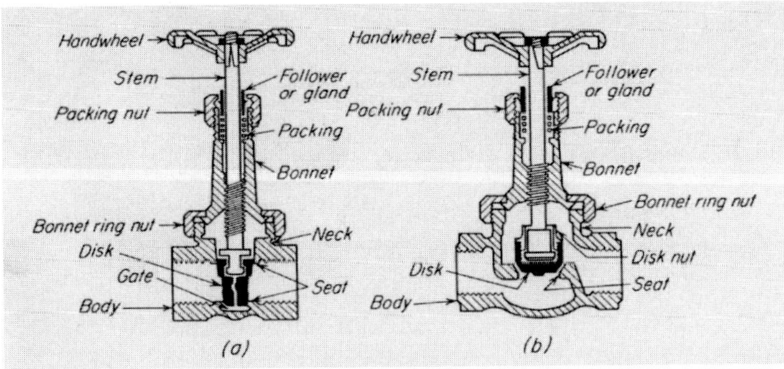

Figure A8-0-2: Control Valve Final Product: (a) Gate Valve. (b) Globe Valve

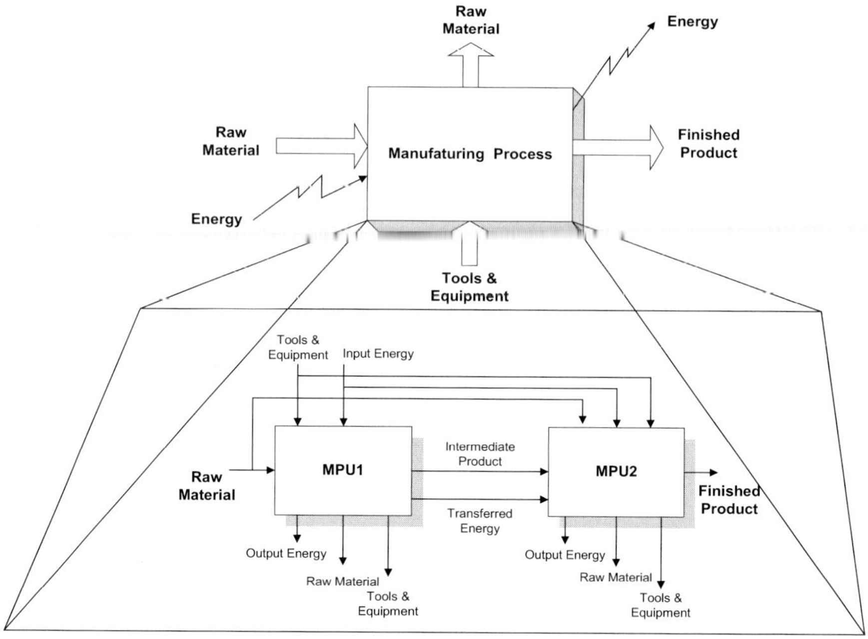

Figure A8-0-3: Manufacturing Process Model in POOM

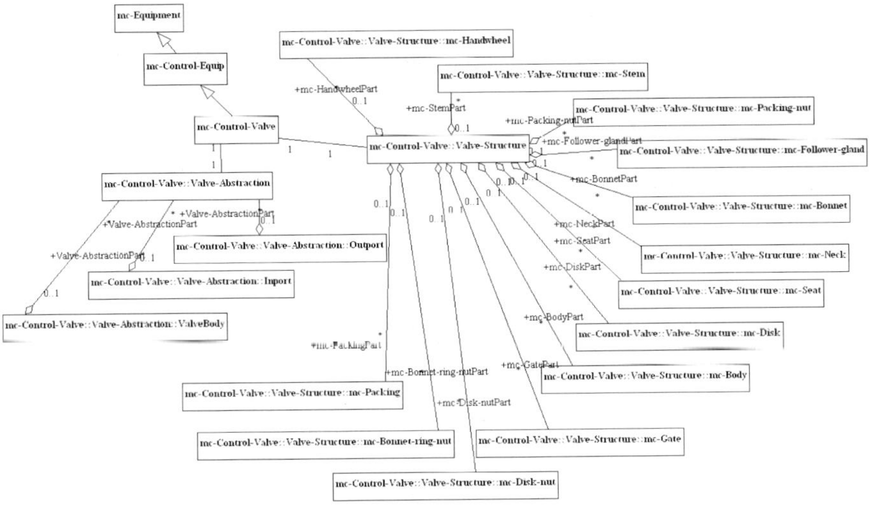

Figure A8-0-4: Object-Oriented Model Representation of Control Valve

Table A8-1: Simplified Table Describing the Statechart of Product Object

State	Event	Actions/Activities (Operation)
IP1	Event1 <Initial product is entered>	Operation11 <cut edge “E1”> Operation12 <make hole “H1”>
IP2	Event2 <Intermediate product IP1 is entered>	Operation21 <connect screw “S1”> Operation22 <cover by silver “L1”>

Table A8-2: Object Types vs. Action rule sets of the Manufacturing Process of Control Valve

Object Type	Event Type	Action rule set
Product	Initial_Product_entered_into_MPU1	Cut_Edge Make_Hole
Equipment	Installed	Check_Condition
Storage	Start_StockTake	Start_Count

Table A8-3: Action Rule Example for Operation within MPU1

When Initial_Product_entered_into_MPU1	Event type
If Equipment_is_ready	Condition on Object type
Then Cut_Edge	Action rule set
For Initial_Product	Object type

The above tables represent aspects of the manufacturing process modeling using object-oriented approach. Figure A8-5 shows the manufacturing process as represented using class diagram in UML format. In this figure, the input IP0 will be converted into the final product in different steps in different manufacturing process units (MPUs).

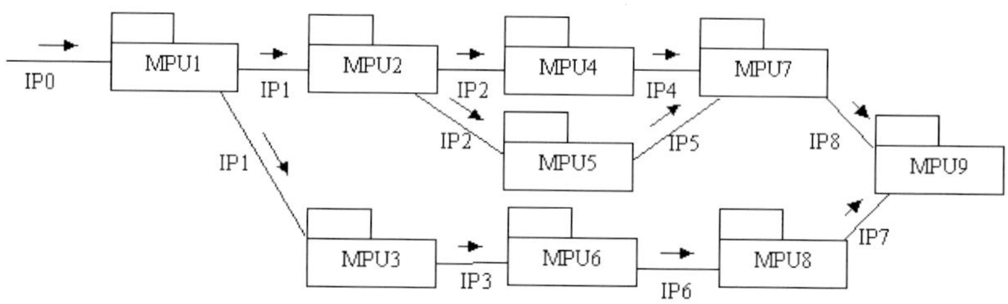

Figure A8-0-5: Simplified Process Diagram for the Manufacturing Process

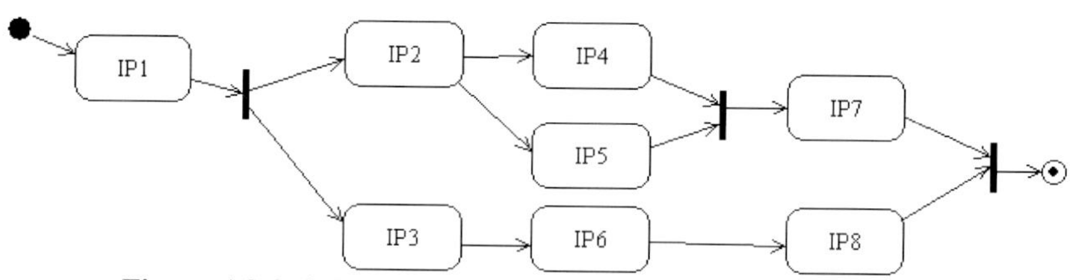

Figure A8-0-6: Manufacturing Process State Diagram

In figure A8-6, the state diagram of the manufacturing process has been presented where each state represents product while it is in the form of an intermediate product. This example can be used to show the supply chain relationship where the manufacturer of the control valve will be supplying the control valves to the chemical plants. This requires the model of the control valve to be supplied as well. This means that the control valve model will include its design intend while operating the plant. This can help in hazard evaluation, fault detection, and plant design i.e. rationales.

Table of Index

O

P

R

S

V